SpringerBriefs in Computer Science

Series Editors
Stan Zdonik
Peng Ning
Shashi Shekhar
Jonathan Katz
Xindong Wu
Lakhmi C Jain
David Padua
Xuemin Shen
Borko Furht
VS Subrahmanian
Martial Hebert
Katsuchi Ikeuchi
Bruno Siciliano

For further volumes:
http://www.springer.com/series/10028

Ben Juurlink • Mauricio Alvarez-Mesa
Chi Ching Chi • Arnaldo Azevedo
Cor Meenderinck • Alex Ramirez

Scalable Parallel Programming Applied to H.264/AVC Decoding

 Springer

Ben Juurlink
Technische Universität Berlin
Berlin, Germany
juurlink@cs.tu-berlin.de

Mauricio Alvarez-Mesa
Technische Universität Berlin
Fraunhofer HHI.
Berlin, Germany
alvarez@ac.upc.edu

Chi Ching Chi
Technische Universität Berlin
Berlin, Germany

Arnaldo Azevedo
Delft University of Technology
Delft, The Netherlands

Cor Meenderinck
IntelliMagic. Leiden
The Netherlands

Alex Ramirez
Universitat Politècnica de Catalunya
Barcelona Supercomputing Center
Barcelona, Spain

ISSN 2191-5768   ISSN 2191-5776 (electronic)
ISBN 978-1-4614-2229-7 ISBN 978-1-4614-2230-3 (eBook)
DOI 10.1007/978-1-4614-2230-3
Springer New York Heidelberg Dordrecht London

Library of Congress Control Number: 2012938716

Printed on acid-free paper

Springer is part of Springer Science+Business Media (www.springer.com)

What is scalability?

Title of an article authored by Mark Hill that appeared in SIGARCH Comput. Arch. News, 18(4), Dec. 1990.

Preface

Welcome to this book! In it we share our experiences in developing highly efficient and scalable parallel implementations of H.264/AVC video decoding, a state-of-the-art video coding standard. We hope to convince you, the reader, that scalable parallel programming is an art and that substantial progress is still needed to make it feasible for non-expert programmers. We also present a parallel-application-design-process and hope that this design process will make the development of parallel applications easier.

When we were invited by Springer to write a SpringerBrief book because our article "Parallel Scalability of Video Decoders" was among the top-downloaded articles from the Journal of Signal Processing Systems, we accepted the invitation but did not want to simply extend the original article. Instead, we decided to present all or at least most of the work we did on this very exciting topic in order to be able to present a complete picture. This was easier said than done. For one, the work was done over a time span of several years and in several years the state-of-the-art computing systems change dramatically. Because of this, the computing systems used to evaluate the presented implementations may differ in each chapter. Please forgive us for these inconsistencies. If we had had more time and money, we would have done it differently.

This book is targeted at graduate students, teachers in higher education, and professionals who would like to understand what it means to parallelize a real application. While there are many textbooks on parallel programming and parallel algorithms, for understandable reasons of space, very few discuss real applications. It is also targeted at video coding experts who know a lot about video coding but who would like to know how it could be parallelized and which features of modern multi-/many-core architectures need to be exploited in order to develop efficient implementations. When reading this book they will probably smile because we use some of their terms wrongly or in a different context than they usually do. Well, we will smile back at them when they confuse an SMP with a cc-NUMA or vice versa.

This book may be used in several ways. For example, it may be used as a supplement to a parallel programming, parallel computer architecture, or parallel algorithms course. One of the authors uses this material to give two lectures about "The

Art of Parallel Programming - H.264 Decoding as a Case Study" at the end of a course on multicore systems (slides including exercises are available on request). This was also the title we originally had in mind for this book, but Springer thought that this title was too long. We accepted this advice as they are, after all, in the publishing business.

While reviewing this book one of the authors mentioned that we are too negative, that this book might scare away people from parallel programming. This is certainly not our intention, and we have revised the book to make it more positive. On the contrary, by presenting a parallel-application-design-process we hope to interest more people in the art of parallel programming.

This book was mainly written by the first three authors. The other three authors, however, contributed significantly to the articles on which this book is based, and therefore they are rightfully mentioned as co-authors. We especially would like to thank our families (Claudia, Lukas, Leon, Claudia, Luna, Ozana, Alex, Marieke, Alicia, Helena, Martí) for their love and support. This book had to be written partially in our spare times, which should be devoted to them. We would also like to thank Senj Temple, the first author's Canadian sister-in-law, for proofreading several parts of this book. None of the authors is a native English speaker and her feedback was really helpful. Also thanks to Biao Wang for his help with the encoding of some of the videos used in this book and for providing a nice IDCT example.

Berlin,   *Ben Juurlink, Mauricio Alvarez-Mesa,*
March 2012   *Chi Ching Chi*

Acknowledgements

This work described in this book was supported in part by the European Commission in the context of the SARC integrated project, grant agreement no. 27648 (FP6), the ENCORE project, grant agreement no. 248647 (FP7), and the European Network of Excellence on High-Performance Embedded Architecture and Compilation (HiPEAC).

Contents

1 Introduction 1
References 3

2 Understanding the Application: An Overview of the H.264 Standard . 5
2.1 Introduction 5
2.2 High-level Overview 6
2.3 Elements of a Video Sequence 8
2.4 Frame Types 9
2.5 H.264 Coding Tools 9
  2.5.1 Entropy Coding 9
  2.5.2 Integer Transform 10
  2.5.3 Quantization 11
  2.5.4 Inter-Prediction 11
  2.5.5 Intra-Prediction 12
  2.5.6 Deblocking filter 13
  2.5.7 Comparison With Previous Standards 13
2.6 Profiles and Levels 13
2.7 Conclusions 15
References 15

3 Discovering the Parallelism: Task-level Parallelism in H.264 Decoding 17
3.1 Introduction 17
3.2 Function-level Decomposition 18
3.3 Data-level Decomposition 19
  3.3.1 Frame-level Parallelism 19
  3.3.2 Slice-level Parallelism 20
  3.3.3 Macroblock-level Parallelism 21
  3.3.4 Other Data-level Decompositions 31
3.4 Conclusions 32
References 33

4   Exploiting Parallelism: the 2D-Wave 35
   4.1   Introduction 35
   4.2   Cell Architecture Overview 36
   4.3   Task Pool Implementation 37
   4.4   Ring-Line Implementation 41
   4.5   Experimental Evaluation 46
        4.5.1   Performance and Scalability 46
        4.5.2   Profiling Analysis 48
   4.6   Conclusions 51
   References 52

5   Extracting More Parallelism: the 3D-Wave 53
   5.1   Introduction 53
   5.2   Dynamic 3D-Wave Algorithm 54
   5.3   2D-Wave Implementation on a Shared-Memory System 55
   5.4   Dynamic 3D-Wave Implementation 58
   5.5   Experimental Evaluation 60
        5.5.1   Experimental Setup 60
        5.5.2   Experimental Results 62
   5.6   Conclusions 65
   References 66

6   Addressing the Bottleneck: Parallel Entropy Decoding 67
   6.1   Introduction 67
   6.2   Profiling and Amdahl 68
   6.3   Parallelizing CABAC Entropy Decoding 71
        6.3.1   High-Level Overview of CABAC 72
        6.3.2   Frame-level Parallelization of CABAC 73
   6.4   Experimental Evaluation 75
        6.4.1   Experimental Setup 75
        6.4.2   Experimental Results 76
   6.5   Conclusions 78
   References 78

7   Putting It All Together: A Fully Parallel and Efficient H.264 Decoder 81
   7.1   Introduction 81
   7.2   Pipelining H.264 Decoding 82
   7.3   Parallel Entropy Decoding 84
   7.4   Parallel Macroblock Reconstruction 85
   7.5   Dynamic Load Balancing using Unified Decoding Threads 86
   7.6   Experimental Evaluation 91
   7.7   Conclusions 95
   References 96

8   Conclusions 97
   References 101

Acronyms

AVC	Advanced Video Coding
CABAC	Context Adaptive Binary Arithmetic Coding
CAVLC	Context Adaptive Variable Length Coding
CRF	Constant Rate Factor
DCT	Discrete Cosine Transform
DLP	Data-level Parallelism
DMA	Direct Memory Access
DPB	Decoded Picture Buffer
EDT	Entropy Decoding Thread
EIB	Element Interconnect Bus
GOP	Group of Pictures
HEVC	High Efficiency Video Coding
MB	Macroblock
MPEG	Moving Pictures Expert Group
MRT	Macroblock Reconstruction Thread
MV	Motion Vector
MVP	Motion Vector Prediction
NUMA	Non-Uniform Memory Architecture
ORL	Overlapping Ring-Line
PPE	Power Processing Element
QP	Quantization Parameter
RL	Ring-Line
SIMD	Single Instruction Multiple Data
SPE	Synergistic Processing Elements
SPMD	Single Program Multiple Data
SSB	Slice Syntax Buffer
TLP	Thread-level Parallelism
TP	Task Pool
UDT	Unified Decoding Thread
VCEG	Video Coding Experts Group

CPM
...
Cricket
...
WM
...
...
CPE
...
MPS/Cite
...
...
LMF
...
JDBC
...
DBE
...
...
CET
...
EBS
...
...
...
...
MBI
...
M-PES
...
ORE
...
GM
...
SNTM-DB
...
LOM
...
CIT
...
OLE
...
GPL
...
Co.
...
CHQ
...
SOA
...
BS/TU
...
HDP
...
TD
...
HD
...
SWDB
...

Chapter 1
Introduction

Dear reader,

We live in turbulent times. We are not talking about some worldwide crisis, but about the *parallel software crisis*. Some online articles even warn us that the free lunch is over [4]. What has happened?

For several decades, most software developers enjoyed a free lunch. If their applications were not feasible given the performance of contemporary computer systems, they just had to be patient and wait a few years until computers were powerful enough to execute their applications. This comfortable situation was enabled by *Moore's law*, which states that the number of transistors that can be placed on a chip approximately doubles every two years. While increasing the number of transistors does not necessarily mean that processors become faster, microelectronic engineers and computer architects have succeeded in increasing their performance at roughly the same rate. They exploited the additional transistors as well as the fact that transistors have become faster in order to invent techniques such as pipelining, out-of-order and superscalar execution, branch prediction, and speculative execution to keep increasing performance.

By plus or minus 2005, however, this comfortable situation suddenly ended. The performance of a single processing unit, or *cores* as they are now called, stopped increasing. But Moore's law was and is still valid, and major processor vendors such as Intel, AMD, and IBM started to use the extra transistors to place several cores on a single chip. A processor now can contain several cores and we have entered the era of *multicores*. A commodity processor in a PC now contains 4 to 8 cores, and processors with 50+ cores, such as Intel's Many Integrated Core (MIC) [2] or Tilera's TILE-Gx [1], are available in labs or have been announced. What has caused this shift?

Every time a transistor switches (and even when it is idle), a small amount of power is dissipated and some energy is lost in the form of heat. While very tiny, the sheer number of transistors on a chip make the chip dissipate significant amounts of power and generate significant amounts of heat, and therefore a cooling system (fan) is required. The *power density* (the amount of power per square mm), however, is limited, or otherwise the chip is damaged. Since power increases roughly quadratically with clock frequency, it is more power-efficient to keep the clock frequency constant but to increase the number of cores instead. As a result, the frequency of the last generation of processors has remained constant (in some cases it has decreased). A recent estimation made by the International Technology Roadmap for Semiconductors is a frequency increase of *at most* 1.25X per technology generation [3].

But how does this affect software development and take away our free lunch? Well, in order to enjoy the performance benefits that multi-/many-core systems have to offer, software developers have to *explicitly parallelize* their applications. Performance benefits no longer come for free. Furthermore, to prevent having to redesign the application every time a new processor with more cores is released, applications need to be parallelized in a *scalable* way. There is no precise definition of scalability, but informally it means that an application is able to increase its performance when resources (i.e., cores) are added. Developing scalable parallel applications, however, is considered by many to be a very complex task. Indeed, for many thousands of years humanity has tended to think sequentially by, e.g., putting our socks on before our shoes, and by writing letter for letter and word for word. At least the male part of humanity tends to think sequentially, women seem to be more capable of multi-tasking.

But what makes parallel programming so much more difficult than sequential programming? One of the authors claimed the complexity of parallel programming is exponentially larger than the complexity of sequential programming. But is this really true? And if this is the case, what can be done to reduce the complexity? In this book we try to provide answers to these questions by sharing our experiences in developing highly efficient and scalable parallel implementations of H.264/AVC decoding. The answers are necessarily short and concise, because this book has a page limit of 100 pages (otherwise, it would not be called a "brief"). H.264/AVC decoding is a challenging and hence interesting application, because it is highly irregular and very dynamic. Furthermore, there is no single kernel that consumes 99.9% of the execution time and scalable parallel implementations need to exploit pipelining as well as data-level parallelism (please do not worry, we will explain these terms later).

One contribution of this book is a parallel-application-design-process consisting of a number of steps. This book is organized according to these steps (we have to admit that the steps were identified after we had done most of the work). The first step is to *understand the application*. Therefore, in Chapter 2 we present a brief overview of the H.264/AVC standard. The second step is to *discover the parallelism* in the application. Chapter 3, therefore, discusses several ways or approaches to parallelize H.264 decoding. The two most promising approaches, referred to as the 2D-Wave and the 3D-Wave, are analyzed in detail using an ideal model that ignores communication and synchronization cost. Finding the parallelism, however, is not sufficient. Application developers also have to find ways to efficiently implement the algorithm, i.e., to *map the parallelism onto the architecture*. In Chapter 4 we take the 2D-Wave algorithm and present how it can be efficiently mapped onto a multi-core architecture. To be precise, we present and compare two implementations: one that optimally distributes work via a so-called task pool and one that statically assigns work to cores. The 2D-Wave, however, scales efficiently to only 20-25 cores for Full High Definition (FHD) resolution videos. While this is more than sufficient for real-time execution of current video resolutions, academic curiosity as well as the requirements of future video resolutions motivated us to determine the scalability of H.264 decoding to many-core processors. Therefore, the next step that application developers have to take if their application requires higher performance is to *extract more parallelism* from the application. In Chapter 5 we illustrate this step by presenting an implementation of the 3D-Wave algorithm, which exploits parallelism within frames (as the 2D-Wave) as well as between frames and, consequently, achieves much higher scalability than the 2D-Wave. As will be shown, however, implementing the 3D-Wave is quite intricate because the dependencies between tasks are not known a priori. Chapter 4 and Chapter 5 focus on the macroblock (MB) reconstruction phase of H.264 decoding, which is the most time consuming phase of H.264 decoding. There is another phase, the entropy decoding phase, that consumes a significant amount of time, but this phase cannot be parallelized in a way similar to the MB reconstruction phase. Basically, in Chapter 4 and Chapter 5 we assume that the entropy decoding phase is performed by a hardware accelerator and, indeed, commercial H.264 decoders typically contain a hardware IP block to perform entropy decoding. But for higher resolution and higher bitrates it is necessary to parallelize entropy decoding and in Chapter 6 we describe how it can be done. This step is called *addressing the bottleneck* or, jestingly, *defeating Amdahl's law*. The final step is to put all of the work together in order to produce a highly efficient and scalable application that includes all stages. We call this step *re-iterate, do it all over again*, or *put it all together*. In Chapter 7 this step is demonstrated by presenting a highly efficient parallel H.264 video decoder that is capable of decoding Quad High Definition (QHD) (researchers from the video coding domain appear to say "2160p50") resolution videos in (faster than) real time. It exploits pipelining as well as data-level parallelism and is scalable to a large number of cores. Finally, Chapter 8 briefly summarizes each chapter, presents our final conclusions, and briefly describes our work on parallelizing the upcoming High Efficiency Video Coding (HEVC) standard, which is currently under development. Different from H.264, where parallelism was an afterthought, HEVC includes proposals specifically aimed at making it amenable for parallelization.

References

1. Bell, S., Edwards, B., Amann, J., Conlin, R., Joyce, K., Leung, V., MacKay, J., Reif, M., Bao, L., Brown, J., Mattina, M., Miao, C.C., Ramey, C., Wentzlaff, D., Anderson, W., Berger, E., Fairbanks, N., Khan, D., Montenegro, F., Stickney, J., Zook, J.: TILE64 - Processor: A 64-Core SoC with Mesh Interconnect. In: Proceedings of the IEEE International Solid-State Circuits Conference, ISSCC, pp. 88 –598 (2008)
2. Heinecke, A., Klemm, M., Bungartz, H.J.: From GPGPU to Many-Core: Nvidia Fermi and Intel Many Integrated Core Architecture. Computing in Science Engineering 14(2), 78 –83 (2012)
3. ITRS: International Technology Roadmap for Semiconductors 2011 update system drivers (2011). URL http://www.itrs.net/Links/2011ITRS/2011Chapters/

4. Sutter, H.: The Free Lunch is Over: A Fundamental Turn Toward Concurrency in Software. Dr. Dobb's Journal **30**(3) (2005). URL `http://www.gotw.ca/publications/concurrency-ddj.htm`

Chapter 2
Understanding the Application: An Overview of the H.264 Standard

Abstract Before any attempt to parallelize an application can be made, it is necessary to understand the application. Therefore, in this chapter we present a brief overview of the state-of-the-art H.264/AVC video coding standard. The H.264/AVC standard is based on the same hybrid structure as previous standards, but contains several new coding tools that increase the coding efficiency and quality. These new features increase the computational complexity of video encoding as well as decoding, however. Therefore, parallelism is a solution to obtain the performance required for real-time processing. The goal of this chapter is not to provide a detailed overview of H.264/AVC, but to provide sufficient background to be able to understand the remaining chapters.

2.1 Introduction

A video consists of a sequence of pictures or *frames*. When these frames are displayed rapidly in succession and provided the *frame rate* is high enough (typically 20-25 frames per second), viewers have the illusion that motion is occurring. To be able to store and transmit large video sequences, they need to be compressed or *encoded*. For example, to store a 2-hour uncompressed video of Full High Definition (HD) resolution frames (the resolution refers to the frame size and FHD corresponds to a frame size of 1920×1080 picture elements or *pixels*), a capacity of 2 (hours) $\times$ 60 (minutes per hour) $\times$ 60 (seconds per minute) $\times$ 25 (frame rate, frames per second) $\times$ 1920×1080 (frame size in pixels) $\times$ 3/2 (number of bytes per pixel) = 559,9 GB is required, which far exceeds the capacity of current optical disc storage (50 GB for dual layer Blu-ray Disc). Therefore, videos have to be encoded before they can be stored or transmitted and decoded before they can be displayed.

To encode a video sequence, *spatial* and *temporal redundancy* can be exploited. Spatial redundancy means that pixels that are spatially close to each other typically have similar values. For example, if one pixel is "green", then its neighboring pixels are often also green or slightly lighter or darker green. This redundancy can be

5

exploited to compress the pixel values in fewer bits. Another form of redundancy is temporal redundancy. Temporal redundancy refers to the fact that consecutive frames are typically very similar. Very often, consecutive frames are almost identical except that they have shifted a little due to motion. This redundancy can be exploited by storing the difference or *error* between a frame and a previous or next frame corrected by the amount of shift. Since the differences are usually small, they can be encoded in fewer bits than the original pixel values. This process is known as *motion compensation* and the amount of shift is given by the *motion vector*.

Currently, the best video coding standard, in terms of compression rate and quality, is H.264/AVC (Advanced Video Coding) [1], also known as MPEG-4 part 10[1]. It is used in Blu-ray Disc and many countries are using or will use it for terrestrial television broadcast, satellite broadcast, and mobile television services. It improves the compression ratio of previous standards such as H.262/MPEG-2 and MPEG-4 Advanced Simple Profile (ASP) by a factor of 2 or more. The H.264 standard [13] was designed to serve a broad range of application domains ranging from low to high bitrates, from low to high resolutions, and a variety of networks and systems, e.g., Internet streams, mobile streams, disc storage, and broadcast. It was jointly developed by ITU-T Video Coding Experts Group (VCEG) and ISO/IEC Moving Picture Experts Group (MPEG).

H.264 is based on the same block-based motion compensation and transform-based coding framework as prior MPEG and ITU-T video coding standards [10]. It provides higher coding efficiency by adding features and functionality that, in turn, increase the computational complexity. H.264 includes many new coding tools for different application scenarios but in this chapter only a summary of these tools can be presented, focusing on the tools which effect the performance of the decoder, which is the focus of this book, the most.

2.2 High-level Overview

The input of the video decoder is a compressed video, also called a *bitstream*. After a series of steps, referred as video decoding *kernels*, the decoder produces a reconstructed video than can be displayed on a screen. Real-time is achieved when each frame is processed within a limited amount of time, and the required frame rate is commonly expressed in terms of frames per second (fps).

A high-level view of the H.264 decoder is presented in Figure 2.1. The decoder is composed of four main stages: entropy decoding, inverse quantization and inverse discrete cosine transform (ICDT), prediction, and the deblocking filter.

First, the compressed bitstream has to be *entropy decoded*. Entropy coding is a lossless data compression scheme that replaces each fixed-length input symbol by a variable-length code word. Since shorter code words are used for more frequent symbols, this compresses the input data. Entropy decoding decompresses the bit-

---

[1] For brevity, in the remainder of this book, H.264/AVC will be referred to as H.264

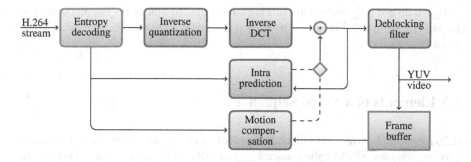

Fig. 2.1: High-level block diagram of the H.264 decoder.

stream and produces a series of syntax elements that specify, e.g., the coding modes, the quantized coefficients, motion vectors, etc. These elements are used by the other stages to reconstruct the final output video.

After entropy decoding, an inverse quantization step is applied to the quantized coefficients. At the encoder side, the coefficients are quantized by dividing them by a value greater than one. Again, the goal is to make the values smaller so that they can be encoded in fewer bits. Quantization is the main reason why video coding is *lossy*, i.e., some information is lost. Usually, however, the error is too small to be noticeable for the human eye. Then, an inverse DCT (IDCT) is applied to the reconstructed coefficients to obtain the residual data. The residual data represent the difference between the original signal and the predicted signal. The DCT, which is performed by the decoder, concentrates the signal information in a few components, which again compresses the information, and the decoder has to perform the inverse transform.

The predicted signal is created either with *intra-* or *inter-prediction*. In intra-prediction the predicted signal is formed using only samples from the same picture. Typically, the first frame from a scene is encoded using intra-prediction, since there is no previous frame very similar to this one. With inter-prediction, the signal is predicted from data of other frames, called *reference frames*, using *motion compensation*. Basically, the encoder partitions each frame into square blocks called *macroblocks* (MBs). For each MB it searches for similar blocks (not necessarily MBs) in the reference frames and it encodes the difference between the MB and the blocks that are similar (we do not have to repeat that the goal is to produce smaller values that can be encoded in fewer bits). The thus obtained motion vectors are included in the bitstream and the decoder uses this information to predict the signal. Thereafter, the decoder adds the predicted signal to the residual data to obtain the reconstructed signal.

Partitioning each frame into MBs, however, can lead to visible quality losses known as *blocking artifacts*. To improve the visual quality of the final output signal, a *deblocking filter* is finally applied to the reconstructed signal.

If you do not fully understand this high-level description of H.264, please do not despair. In Section 2.5 we describe each phase, or *coding tool*, in somewhat more detail.

2.3 Elements of a Video Sequence

H.264 is a *block-based* coder/decoder (codec), meaning that each frame is divided into small square blocks called macroblocks (MBs). The coding tools / kernels are applied to MBs rather than to whole frames, thereby reducing the computational complexity and improving the accuracy of motion prediction. Figure 2.2 depicts a generic view of the data elements in a video sequence. It starts with the sequence of frames that comprise the whole video. Several frames can form a Group of Pictures (GOP), which is an independent set of frames. Each frame can be composed of independent sections called *slices*, and slices ones, in turn, consist of MBs. Each MB can be further divided into sub-blocks, which in turn, consist of pixels.

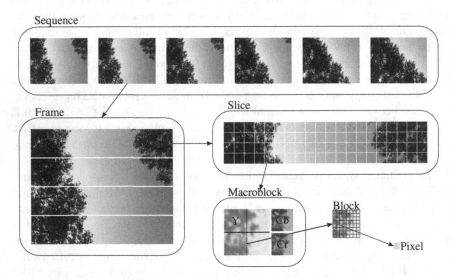

Fig. 2.2: Elements of a video sequence

A MB[2] consists of separated blocks for luma (denoted by Y) and chroma signals (denoted by Cb and Cr). A pre-processing step has to be applied to convert video from a different color component format (such as red-green-blue, RGB) to the YCbCr color model. Chroma sub-sampling is applied to reduce the amount of

---

[2] The word macroblock is used so often in this book that we introduce an acronym for it: MB. When reading the text, however, we read it as macroblock. Therefore, we write "a MB", while it should be written as "an MB" when the acronym is pronounced as "em-bee".

color information, since the human eye is more sensitive to brightness (Y) than to color (Cb and Cr) [8]. The most common color structure is denoted by 4:2:0 in which the chroma signals (Cb and Cr) are sub-sampled by 2 in both dimensions. As a result, in H.264, as in most MPEG and ITU-T video codecs, each MB typically consists of one 16×16 luma block and two 8×8 chroma blocks.

2.4 Frame Types

H.264 defines three main types of frames: *I-*, *P-*, and *B-frames*. An I-frame uses intra-prediction and is independent of other frames. In intra-prediction, each MB is predicted based on adjacent blocks from the same frame. A P-frame (Predicted frame) uses motion estimation as well as intra-prediction and depends on one or more previous frames, which can be either I-, P- or B-frames. Motion estimation is used to exploit temporal correlation between frames. Finally, B-frames (Bidirectionally predicted frames) use bidirectional motion estimation and can depend on previous frames as well as future frames [3].

Figure 2.3 illustrates a typical I-P-B-B (first an I-frame, then two B-frames between P-frames) sequence. The arrows indicate the dependencies between frames caused by motion estimation. In order to ensure that a reference frame is decoded before the frames that depend on it, and because B-frames can depend on future frames, the decoding order (the order in which frames are stored in the bitstream) differs from the display order. Thus a reordering step is necessary before the frames can be displayed, adding to the complexity of H.264 decoding.

2.5 H.264 Coding Tools

The H.264 standard has many coding tools each one with several options. Here we can only briefly mention the key features and compare them to previous standards.

2.5.1 Entropy Coding

H.264 includes two different entropy coding techniques. The first one is *Context Adaptive Variable Length Coding* (CAVLC), which is an adaptive variant of Huffman coding and targeted at applications that require a slightly simpler entropy decoder. The second one is *Context Adaptive Binary Arithmetic Coding* (CABAC), which is based on arithmetic coding techniques. Arithmetic coding differs from Huffman coding in that it does not encode each symbol separately, but encodes an entire message in a single number between 0.0 and 1.0. This results in compression ratios that are 10% to 14% higher than CAVLC [6]. In this work we focus on H.264

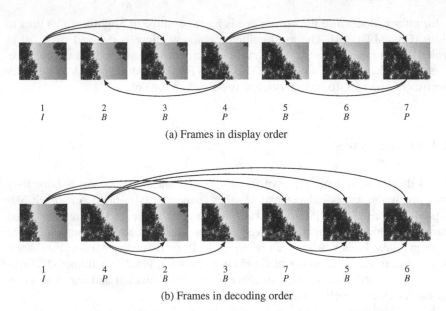

(a) Frames in display order

(b) Frames in decoding order

Fig. 2.3: Types of frames and display order versus decoding order.

decoders that employ CABAC, because it achieves higher compression ratios than CAVLC, and because it is widely used in HD applications.

2.5.2 Integer Transform

The Discrete Cosine Transform (DCT) is a mathematical transform that converts input data (e.g. residual data) into a different domain (called the frequency domain). The main purpose of this conversion is to separate the input data into independent components and concentrate most of the information on a few coefficients [9]. The 2-dimensional Discrete Cosine Transform (2D-DCT) is the most widely used transform for block-based video codecs because its mathematical properties allow efficient implementations [2].

H.264 transform is an integer approximation of the DCT allowing implementations with integer arithmetic for decoders and encoders. The main transform is applied to 4×4 blocks and is useful for reducing ringing artifacts (negative coding effects that appear as bands near the edges of objects in an image). The H.264 High Profile (described later) allows another transform for 8×8 blocks which is useful in HD video for the preservation of fine details and textures [11]. The encoder is allowed to select between the 4×4 and 8×8 transforms on a macroblock by macroblock basis. The transforms employ only integer arithmetic without multiplications, with coefficients and scaling factors that allow for 16-bit arithmetic computation [5]. In the decoder, an Inverse Discrete Cosine Transform (IDCT) is applied to the transformed coefficients in order to reconstruct the original residual data.

Matrices 2.1 and 2.2 illustrate a 8×8 IDCT using data from a real H.264 video. X represents a block with input transformed coefficients, most of them with zero value:

$$X = \begin{bmatrix} -936 & -1020 & -460 & 0 & 0 & 68 & 0 & 0 \\ 0 & 128 & 344 & 128 & 0 & 0 & 0 & 0 \\ 276 & 86 & 232 & 0 & 0 & 0 & 0 & 0 \\ 136 & 64 & 0 & 0 & 0 & 0 & 0 & 0 \\ 216 & 0 & 0 & 0 & 0 & 0 & 0 & 0 \\ 0 & 0 & 0 & 0 & 0 & 0 & 0 & 0 \\ 0 & 0 & 0 & 0 & 0 & 0 & 0 & 0 \\ 0 & 0 & 0 & 0 & 0 & 0 & 0 & 0 \end{bmatrix} \quad (2.1)$$

After applying the IDCT the reconstructed version of the coefficients Y is obtained:

$$Y = \begin{bmatrix} -11 & -18 & -18 & -14 & -4 & 5 & 14 & 11 \\ -31 & -35 & -31 & -23 & -12 & -4 & 3 & 0 \\ -51 & -49 & -35 & -24 & -12 & -7 & -3 & -6 \\ -50 & -45 & -26 & -14 & -3 & 1 & 2 & -2 \\ -53 & -43 & -19 & -6 & 3 & 4 & 3 & -1 \\ -56 & -45 & -20 & -8 & 0 & -1 & -2 & -4 \\ -56 & -43 & -18 & -6 & 0 & 0 & 1 & 1 \\ -52 & -39 & -13 & -1 & 6 & 5 & 9 & 10 \end{bmatrix} \quad (2.2)$$

2.5.3 Quantization

The encoder quantizes the transformed coefficients by dividing them by values greater than 1, so the quantized coefficients can be coded using fewer bits. The decoder has to perform the inverse operation, i.e., multiply the quantized coefficients with the same values. Quantization is the main reason why video coding is *lossy*, meaning that some information is lost due to rounding errors. The loss of information is, however, usually too small for humans to notice.

2.5.4 Inter-Prediction

Advances in inter-prediction (predicting a MB from one or several MBs in other frame(s)) is one of the main contributors to the compression improvement of H.264. The standard allows variable block sizes ranging from 16×16 down to 4×4, and each block has its own motion vector(s). The motion vector is quarter sample accurate, meaning that motion vectors can also point to positions between pixels and the best matching block is obtained using interpolation. Multiple reference frames can be used in a weighted fashion [3]. Using multiple reference frames significantly improves coding *occlusion areas*, where an accurate prediction can only be made from a frame further in the past. Figure 2.4 illustrates motion estimation.

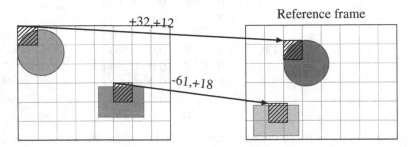

Fig. 2.4: Motion estimation. For each MB the best matching block in the reference frame is found. The encoder codes the differences (errors) between the MBs and their best matching blocks. Arrows indicate motion vectors and are labeled by the vector coordinates. In this example the shapes are identical but their colors are slightly larger/darker.

2.5.5 Intra-Prediction

The first frame in a new scene is typically not similar to a previous frame. Therefore, these frames are often I-frames which are not predicted from previous or next frames but using intra-prediction, meaning from itself. P-frames employ inter- as well as intra-prediction. H.264 supports three types of intra coding, denoted by Intra_4x4, Intra_8x8, and Intra_16x16.

Intra_4x4 is well suited for coding parts of a picture with significant spatial detail. The 16×16-pixel MB is divided into sixteen 4×4 sub-blocks and intra-prediction is applied to each sub-block. The standard defines nine prediction modes that the encoder can choose independently for each sub-block. Samples are predicted using previously decoded samples from the blocks to the north, north-east, and west.

The High Profile defines intra-prediction for 8×8 blocks. The prediction modes are basically the same as in 4×4 intra-prediction with the addition of low-pass filtering to improve prediction performance [11].

Intra_16x16 predicts a whole MB and is well suited for coding frame areas with smooth variations of tone and color. Four different prediction modes are available for this type of prediction: vertical, horizontal, DC-, and plane-prediction. The first three ones are very similar to the modes available for 4×4 blocks. Plane-prediction employs a linear function between neighboring samples [13, 7].

2.5.6 Deblocking filter

Processing a frame in MBs can produce artifacts at the block edges, generally considered the most visible artifact in standards prior to H.264. These *blocking artifacts* can be reduced so that they are no longer visible by applying a deblocking filter to the pixels around the block edges, which basically smooths the block transitions. The filter strength is adaptable through several syntax elements [4]. In H.263+ this feature was optional, but in H.264 it is standard, significantly improving the quality of the produced pictures. Furthermore, the deblocking filter is applied before the frame is used as a reference frame, which improves the efficacy of motion compensation.

2.5.7 Comparison With Previous Standards

In Table 2.1 the main features of H.264 are summarized and compared to the MPEG-2 and MPEG-4 standards [13, 12]. One of the main differences between H.264 and the other video codecs is that H.264 allows a variable block size for motion compensation, while MPEG-2 only supports 16×16 pixel blocks and MPEG-4 16×16 down to 8×8 pixel blocks. Additionally, H.264 employs quarter sample resolution for motion estimation, a feature that is optional in MPEG-4 and not available in MPEG-2. Another important difference is that H.264 supports multiple reference frames for motion compensation, while the other two standards support only one reference frame. As explained before, by doing so H.264 improves the coding of occlusion areas.

Furthermore, H.264 features more than one intra-prediction mode, which results in higher intra-compression than the single DC-prediction of MPEG-2 and the prediction of transformed coefficients of MPEG-4. H.264 also includes a mandatory in-loop deblocking filter that is not available in MPEG-2 and MPEG-4, and is optional in H.263. In addition, H.264 includes a binary arithmetic coder (CABAC) which achieves higher compression than the conventional entropy coders based on Variable Length Coding (VLC) employed in previous standards. Finally, H.264 employs an adaptive transform size (4×4 and 8×8), and an integer transform that is faster than the fractional transforms of previous standards.

2.6 Profiles and Levels

The H.264 standard was designed to suite the requirements of a broad range of video application domains such as video conferencing, mobile video, as well as consumer and high definition broadcast. Each domain, however, is expected to use only a subset of all available options. For this reason *profiles* and *levels* were specified to mark conformance points. Encoders and decoders that conform to the same profile are

Codec	MPEG-2	MPEG-4 Part II ASP	H.264 High
Macroblock size	16×16	16×16	16×16
Block size	8×8	$16 \times 16, 16 \times 8, 8 \times 8$	$16 \times 16, 16 \times 8, 8 \times 16,$ $8 \times 8, 4 \times 8, 8 \times 4, 4 \times 4$
Transform	8×8 DCT	8×8 DCT	$8 \times 8, 4 \times 4$ integer transform
Pel-Accuracy	$1, 1/2$ pel	$1, 1/2, 1/4$ pel	$1, 1/2, 1/4$ pel
Reference frames	One frame	One frame	Multiple frames (up to 16 frames)
Bidirectional prediction	forward/backward	forward/backward	forward/backward forward/forward backward/backward
Intra-prediction	DC-pred.	coeff. pred.	$4 \times 4, 8 \times 8, 16 \times 16$ spatial
Deblocking filter	No	No	Yes
Weighted prediction	No	No	Yes
Entropy Coding	VLC	VLC	CAVLC, CABAC

Table 2.1: Comparison of video coding standards

guaranteed to interoperate correctly. Profiles define sets of coding tools and algo-rithms that can be used, while levels place constraints on the bitstream parameters.

The standard initially defined two profiles, but has since then been extended sev-eral times with additional profiles for new applications, such as intra-only profiles (for video editing), scalable profiles for Scalable Video Coding (SVC), and multi-view profiles for Multiview Video Coding (MVC). Currently, the main profiles are:

- **Baseline Profile (BP):** the simplest profile mainly used for video conferencing and mobile video.
- **Main Profile (MP):** this profile was intended to be used for consumer broadcast and storage applications, but has been overtaken by the high profile.
- **Extended Profile (XP):** this profile is intended for streaming video, and includes special capabilities to improve robustness.
- **High Profile (HiP):** this profile is intended for high definition broadcast and disc storage, and is used in Blu-ray.

Besides HiP there are three other high profiles that support up to 14 bits per sample (normally a sample is 8-bit), 4:2:2 and 4:4:4 sampling, as well as other features [11].

Besides profiles, 16 levels are currently defined which are used for all profiles. A level specifies, for example, the upper limit for the picture size, the decoder pro-cessing rate, the size of the multi-picture buffers, and the video bitrate. Levels have profile independent as well as profile specific parameters.

2.7 Conclusions

In this chapter we have presented a, necessarily brief, overview of the H.264/AVC standard. Hopefully we have convinced you that it is a complex application with many options, compression algorithms, and dependencies between the different parts. Therefore, discovering the parallelism in it is less than obvious. This is the step that will be undertaken in the next chapter.

References

1. Advanced Video Coding for Generic Audiovisual Services, Recommendation H.264 and ISO/IEC 14 496-10 (MPEG-4) AVC, International Telecommunication Union-Telecommun. (ITU-T) and International Standards Organization/International Electrotechnical Commission (ISO/IEC) JTC 1 (2003)
2. Blinn, J.: What's that deal with the DCT? IEEE Computer Graphics and Applications 13(4), 78–83 (1993)
3. Flierl, M., Girod, B.: Generalized B Pictures and the Draft H. 264/AVC Video-Compression Standard. IEEE Transactions on Circuits and Systems for Video Technology 13(7), 587–597 (2003)
4. List, P., Joch, A., Lainema, J., Bjøntegaard, G., Karczewicz, M.: Adaptive Deblocking Filter. IEEE Transactions on Circuits and Systems for Video Technology 13(7), 614–619 (2003)
5. Malvar, H., Hallapuro, A., Karczewicz, M., Kerofsky, L.: Low-Complexity Transform and Quantization in H. 264/AVC. IEEE Transactions on Circuits and Systems for Video Technology 13(7), 598–603 (2003)
6. Marpe, D., Schwarz, H., Wiegand, T.: Context-Based Adaptive Binary Arithmetic Coding in the H.264/AVC Video Compression Standard. IEEE Transactions on Circuits and Systems for Video Technology 13(7), 620–636 (2003)
7. Ostermann, J., Bormans, J., List, P., Marpe, D., Narroschke, M., Pereira, F., Stockhammer, T., Wedi, T.: Video Coding with H.264/AVC: Tools, Performance, and Complexity. IEEE Circuits and Systems Magazine 4(1), 7–28 (2004)
8. Richardson, I.E.G.: Video Codec Design: Developing Image and Video Compression Systems. John Wiley and Sons (2002)
9. Richardson, I.E.G.: H.264 and MPEG-4. Video Compression for Next-generation Multimedia. Wiley, Chichester, England (2004)
10. Sikora, T.: Trends and Perspectives in Image and Video Coding. Proceedings of the IEEE 93(1), 6–17 (2005)
11. Sullivan, G., Topiwala, P., Luthra, A.: The H.264/AVC Advanced Video Coding Standard: Overview and Introduction to the Fidelity Range Extensions. In: Proceedings SPIE Conference on Applications of Digital Image Processing XXVII, pp. 454–474 (2004)
12. Tamhankar, A., Rao, K.: An Overview of H. 264/MPEG-4 Part 10. In: Proceedings 4th EURASIP Conference focused on Video/Image Processing and Multimedia Communications, p. 1 (2003)
13. Wiegand, T., Sullivan, G.J., Bjøntegaard, G., Luthra, A.: Overview of the H.264/AVC Video Coding Standard. IEEE Transactions on Circuits and Systems for Video Technology 13(7), 560–576 (2003)

Chapter 3
Discovering the Parallelism: Task-level Parallelism in H.264 Decoding

Abstract In the previous chapter we have reviewed the H.264/AVC standard and showed that H.264 decoding is a complex application with many options, different kernels, and dependencies between different parts. In this chapter we focus on the next step: finding or discovering the parallelism in the application. We qualitatively compare different approaches to parallelize H.264 decoding and analyze the two most promising approaches in more detail using high-level simulation. How to actually implement the algorithms and how to map the parallelism onto multi-/many-core architectures will be presented in later chapters.

3.1 Introduction

After the first step (understanding the application), the second step has to be taken: to discover the parallelism in the application. In this chapter we illustrate this step for H.264 decoding, which, as was shown in the previous chapter, is a complex and dynamic application with many options, kernels, dependencies between kernels, data dependencies, etc. Discovering the parallelism and, moreover, finding *significant amounts* of parallelism, is a challenge.

As will be shown in this chapter, there are several ways to parallelize H.264 decoding. In this chapter as well as the next we focus on the macroblock reconstruction phase, which is generally the most time-consuming phase of H.264 decoding. Not all parallelization approaches are equally good, however. Many do not scale well, i.e., do not exhibit sufficient parallelism to be able to achieve good speedup on a many-core processor. Others decrease the coding efficiency because they require frames to be partitioned into slices. Yet others require too fine grain tasks. Because some overhead is associated with each task, in a highly efficient and scalable parallel application tasks cannot be very fine grain. Therefore, main criteria for comparing different parallel algorithms are: the degree of parallelism and scalability (i.e. the maximum number of independent tasks that can be extracted), coding efficiency, and task granularity, as well as load balancing and frame latency.

17

The two most promising approaches, both based on exploiting parallelism between independent macroblocks (MBs), are analyzed in more detail using high-level simulation. High-level simulation in this case means that we ignore details such as communication and synchronization overhead (important details, but details nonetheless). It will be shown that the first algorithm, referred to as the 2D-Wave, exhibits *weak scaling*, i.e. is scalable provided the problem size increases with the number of cores. The second algorithm, on the other hand, referred to as the 3D-Wave, exhibits *strong scaling*, meaning that it is scalable even if the problem size does not increase. In the case of video coding increasing the problem size means higher frame resolution and/or higher frame rate (or, e.g., a larger number of views in the case of 3D coding).

This chapter is organized as follows. In Section 3.2 we briefly consider *function-level* or *pipelining parallelism*. Pipelining is a technique well-known in computer architecture (as well as many other fields) in which the execution of every instruction is divided into stages, and different stages of different instructions are executed in parallel. In the case of H.264 decoding, instructions correspond to frames and stages correspond to the different H.264 kernels such as entropy decoding, the IDCT, and the deblocking filter. In Section 3.3 we focus instead on *data-level parallelism* (DLP), i.e., parallelism that is obtained by dividing the data into smaller parts and processing each part in parallel. DLP can only be exploited, however, if the data parts are independent. There are many ways in which DLP can be exploited in H.264 decoding and we will discuss and analyze several of them. Conclusions are drawn in Section 3.4.

3.2 Function-level Decomposition

In order to parallelize an application, the problem needs to be *decomposed* into sub-problems that can solved in parallel and the solutions of these sub-problems need to be combined to produce the overall solution. H.264 decoding can be parallelized either by a *function-level* or a *data-level* decomposition. (We conjecture that these are the only two possible parallel decompositions, but this is a philosophical issue.) In a function-level decomposition, the different functions or kernels that constitute the algorithm are executed simultaneously but on different intermediate data. This is similar or identical to *pipelining*, where some kind of assembly line is divided into a number of steps or stages, and each stage is performed in parallel with the other stages, but on a different, e.g., car. In a data-level decomposition, the input data is divided into different parts and all parts is processed simultaneously. In this section we briefly consider a function-level decomposition of H.264 decoding.

As was shown in Chapter 2 (Figure 2.1 on Page 7), H.264 decoding consists of a sequence of kernels executed consecutively on the compressed input bitstream. Some of these kernels can be executed in parallel, either on different data or in a pipelined fashion. For example, on a 4-core system, one core could perform entropy decoding of Frame n, another core could apply dequantization and inverse transform to Frame $n-1$, the third core could execute the prediction stage for Frame $n-2$, while the fourth core applies the deblocking filter to Frame $n-3$.

A function-level decomposition of H.264 decoding (as well as other applications) has a number of problems, however. First, *load balancing* is difficult because the time taken by each kernel is not known a priori but depends on the input data. For some frames entropy decoding takes longer than the prediction stage, while for others it is the other way around. The throughput of a pipeline is determined by the longest stage, and therefore the pipeline stages should be balanced. This is not entirely true if multiple frames can be buffered between stages and frames can be processed out of order, but in H.264 decoding this is difficult due to the motion compensation dependencies between frames. Another main drawback of a function-level decomposition is *limited scalability*. The maximum speedup of a pipelined implementation over a non-pipelined implementation is equal to the number of stages. In H.264 decoding, however, as well as in many other application domains, the number of stages/kernels is limited, and a function-level decomposition would not achieve the scalability we target in this book.

Based on these disadvantages, we dismiss a function-level decomposition for now, but will reconsider it in Chapter 7 when we address the CABAC bottleneck. After all, people are supposed to become wiser as they get older.

3.3 Data-level Decomposition

In a data-level decomposition the work is divided by partitioning the data into smaller parts and by processing the parts in parallel. In many cases, after the parts have been processed in parallel, the partial solutions need to be combined to the overall solution. In H.264, data can be decomposed at different *levels* of the data structure used to represent a video sequence (see Figure 2.2 on Page 8), such as frames, slices, and macroblocks (MBs). In this section we have a close look at several of these options.

3.3.1 Frame-level Parallelism

As illustrated in Figure 2.3 on Page 10, in an H.264 frame sequence some frames (such as I- and P-frames) are used as reference frames, while others (B-frames) are not. In such a case, the B-frames can be decoded in parallel after their reference frames have been decoded. In order to exploit this parallelism, one core has to parse the frame headers, identify independent frames, and assign them to different cores.

This form of frame-level parallelism, however, exhibits very limited scalability, since typically there are only a few (two to four) B-frames between P-frames. Furthermore, unlike previous video coding standards, in H.264 B-frames can be used as reference frames [3]. If B-frames are also used a reference frames, the decoder based on frame-level parallelism would find no parallelism. A not-so-clever solution to this problem would be to instruct the encoder to not use B-frames as reference frames, but this would increase the bitrate (because B-frames cannot be reference frames, frames have to predicted from perhaps less similar I- or P-frames). Moreover, the encoder and decoder are completely separated (they might even be produced by different suppliers) and there is no way for the decoder to enforce its preferences on the encoder. For these reasons we also, at least partially, dismiss frame-level parallelism.

3.3.2 Slice-level Parallelism

In H.264, as in most current hybrid video coding standards, each frame can be partitioned into one or more *slices*. Slices have been included in the standard to increase robustness in the presence of transmission errors. In case a transmission error occurs, other slices are not affected and the visual quality only briefly degrades. In order to accomplish this, slices in a frame should be completely independent from each other, meaning that a slice cannot use content from other slices for entropy coding, prediction, or filtering operations [9]. Since slices are completely independent, they can be processed independently and in parallel.

Slices, however, and hence slice-level parallelism, have a number of disadvantages. The first one is that the number of slices per frame, and hence the amount of parallelism, is determined by the encoder. Since, as explained above, the encoder and the decoder are oblivious of each other, if the encoder decides to use only one (or a few) slice(s) per frame, the scalability of a decoder based on slice-level parallelism will be very limited. In fact, because slices are optional, most H.264 videos available on the Internet use one slice per frame.

Second, although slices are completely independent, the H.264 deblocking filter can be applied across slice boundaries. This option is selectable by the encoder, but implies that even with an input sequence with multiple slices, the deblocking filter should be applied to consecutive slices in sequential order. This significantly reduces the speedup that could be achieved with slice-level parallelism.

The main disadvantage of using slices is, however, the bitrate increase. Slices incur coding losses because they limit the number of MBs that can be used for motion prediction and intra-prediction, reduce the time the arithmetic coder has for adapting context probabilities, and increase the number of headers and start code prefixes in the bitstream necessary to signal the presence of slices [7].

Figure 3.1 depicts the bitrate increase as a function of the number of slices employed for four different input videos with a resolution of 1080p. Up to four slices, the coding losses are less than 5%, which may be acceptable for high bitrate applications such as Blu-ray playback. More slices, however, result in huge coding losses. The word huge may seem exaggerated here, but one should not forget that the prime goal of a video coding standard is compression, not parallelism. Furthermore, using many slices results in subjective quality degradation if the deblocking filter is not applied across slice boundaries.

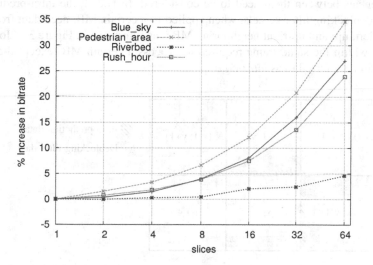

Fig. 3.1: Bitrate increase due to slices for 1080p25 inputs. Fixed Quantization Parameter (QP) equal to 26.

Based on this qualitative analysis, we conclude that slice-level parallelism may be used when the number of slices (and hence cores) is small (e.g. less than or equal to 4) and when the encoder can be controlled. Because of this limited scalability as well as the other limitations, we do not further consider slice-level parallelism in this book, but refer to related work [6, 4] instead.

3.3.3 Macroblock-level Parallelism

"Yet another" data-level decomposition approach is MB-level parallelism that exploits the fact that (some) MBs can be processed in parallel. MBs are not completely independent, however, and any approach based on exploiting MB-level parallelism needs to take the dependencies between MBs into account. In this section we analyze two approaches that exploit MB-level parallelism. The first approach exploits MB-level parallelism within frames and is referred to as the 2D-Wave. The second approach is our novel (at least it was in January 2008 when we first published it) 3D-Wave approach that exploits MB-level parallelism within frames as well as across frames.

3.3.3.1 2D-Wave

In order to be able to exploit the parallelism between different MBs in a frame, the dependencies between them need to be considered. In H.264, the intra-prediction and the deblocking filter kernels, when applied to a certain MB, need data from its left, up-left, up, and up-right neighboring MBs, as illustrated in Figure 3.2. Jointly, the data within the same frame required to process a certain MB (except the MB itself) will be referred to as *intra-data*.

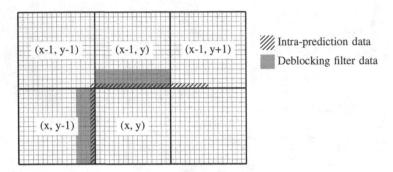

Fig. 3.2: Intra-data of a 16×16 luma block. To intra-predict and deblock $MB(x, y)$, some *decoded* pixels of neighboring MBs are needed. Hashed pixels are required for intra-prediction and colored pixels for the deblocking filter.

When MBs are processed sequentially, they are processed in *scan order*, meaning from left to right and top to bottom. By doing so, the intra-data of a MB is available by the time the MB is decoded (obviously, since this needs to be the case). In other, more scientific sounding words, MB dependencies are preserved when MBs are processed in scan order. The MBs on a "knight-jump-diagonal" are independent, however, and they may be processed in parallel in a "diagonal wavefront manner". As illustrated in Figure 3.3, starting with a MB in the first column, all MBs that can be reached by moving 2 MBs to the right and 1 MB up (knight-jump) are independent and can therefore be decoded simultaneously. In the figure, MBs whose content is fully visible indicate MBs that have been completely processed, dotted MBs correspond to MBs that are currently being decoded in parallel and white MBs correspond to MBs which have not been processed at all. We refer to this high-level parallelization scheme (or approach, or algorithm) as the *2D-Wave*, which is short for two-dimensional wavefront processing. We first read about it in an article of Van der Tol et al. [8], but it might be that someone else discovered it first.

Unlike frame- and slice-level parallelism, the 2D-Wave has the potential to achieve good scalability, since the maximum number of independent MBs is equal to the frame width (in MBs) divided by 2. For 1080p frames (120×68 MBs), this corresponds to 60. Furthermore, the amount of parallelism scales with the resolution. If higher performance is needed because the frame size has increased, the algorithm exhibits more parallelism.

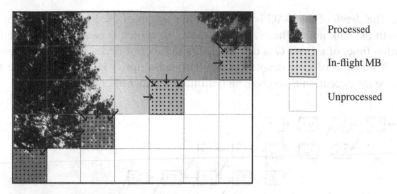

Fig. 3.3: 2D-Wave parallelization: MBs on a diagonal are independent and can be decoded concurrently. The arrows represent the dependencies between MBs.

The 2D-Wave also has some disadvantages, however. First, at the beginning and end of processing a frame, there are only a few independent/parallel MBs. If more cores are employed than there are independent MBs, some cores will have no work to do and remain idle while these MBs are being processed. A second disadvantage is that the entropy decoding phase cannot be parallelized at the MB-level due to bit-level dependencies in the inner-loop of this kernel. (the entropy decoding dependencies will be explained in more detail in Chapter 6). Only after a frame/slice has been entropy decoded, the parallel MB reconstruction phase can start.

High-level Analysis of the 2D-Wave

In the previous section we hopefully have made a convincing argument that the 2D-Wave approach is more scalable than other approaches to parallelize H.264 decoding. But how much parallelism does it really exhibit? In order to answer this question without actually implementing the algorithm, in this section we analyze the algorithm using an analytical approach. In this approach we abstract away details such as communication and synchronization cost (as noted before, important details, but details nonetheless). We are convinced that such a first-order approach is much better than spending several person-months to implement the algorithm only to discover at the end that the amount of parallelism is insufficient to meet the targeted performance.

The parallel processing of MBs during the MB reconstruction phase can be modeled as a *Directed Acyclic Graph* (DAG). Each node in this DAG represents a MB decoding task (to be more precise: a MB reconstruction task, but in this chapter we will neglect this small difference). Graph edges represent data dependencies between the MB decoding tasks; a MB can only be decoded when the MBs on which it depends have been decoded first. The processing of each frame can be represented by a finite DAG. Figure 3.4 depicts the DAG for a 5×5 MBs sample mini-frame. The first MB in the frame ($MB(0,0)$) is the *source node* which has no incoming edges, and the last MB in the frame ($MB(4,4)$) is the *sink node* with no outgoing edges. The *depth* of the DAG is defined as the length of the longest path from the source to the sink node. The *computational work*, which is equal to the sequential execution time, of a DAG G is defined as the number of nodes in G and is denoted by T_s. The depth of G is denoted as T_∞. The subscript ∞ is used to indicate this would be the execution time given an unlimited number of cores.

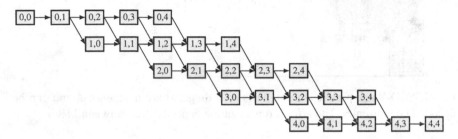

Fig. 3.4: Directed Acyclic Graph (DAG) of MB decoding tasks.

Assuming that a MB decoding task takes constant (unit) time (this is yet another simplification that will be relaxed in a minute) and no synchronization/-communication cost, the maximum speedup can be estimated. Let *mb_width* and *mb_height* respectively denote the width and height of a frame in MBs. Then $T_s = mb\_width * mb\_height$ and $T_\infty = mb\_width + (mb\_height - 1) * 2)$, since, as Figure 3.4 illustrates, the depth of the graph is equal to the number of MB columns plus for every MB row except the first one you have to add 2. Since speedup is defined as the sequential execution time (the amount of work) divided by the parallel execution time (which is at least given by the depth of the DAG), the maximum speedup of the 2D-Wave is given by:

$$Speedup_{max,2D} = \frac{T_s}{T_\infty} = \frac{mb\_width \times mb\_height}{mb\_width + 2 \times (mb\_height - 1)}, \qquad (3.1)$$

Using similar reasoning, the maximum number of MBs that can be processed in parallel can be determined as:

$$ParMB_{max,2D} = \min(\lceil mb\_width/2 \rceil, mb\_height). \qquad (3.2)$$

This equation shows that the number of parallel MBs increases with the frame size. For example, for 1080p (120×68 MBs) frames, the maximum parallelism exhibited by the 2D-Wave approach is $\min(\lceil 120/2 \rceil, 68) = 60$. In Table 3.1, the maximum speedup and maximum number of parallel MBs are given for different video resolutions.

How the number of parallel MBs varies over time can also be determined analytically. Since the MB decoding time is assumed to be constant, we can assign a *time stamp* to each MB, representing the earliest time slot (in MB decoding time units) in which the MB can be decoded. The time stamp of the source node (MB$(0,0)$) is 0, and the time stamp of every other MB is the maximum of the time stamps

Resolution name	Resolution (pixels)	Resolution (MBs)	Total MBs	Max. speedup	Parallel MBs.
UHD, 4320p	7680×4320	480×270	129,600	127	240
QHD, 2160p	3840×2160	240×135	32,400	63.8	120
FHD, 1080p	1920×1080	120×68	8,160	32.1	60
HD, 720p	1280×720	80×45	3,600	21.4	40
SD, 576p	720×576	45×36	1,620	14.1	23

Table 3.1: Maximum speedup of the 2D-Wave algorithm for different video resolutions

of the MBs on which it depends plus 1. We developed a program that builds the DAG for frames of different resolutions, assigns the time stamps, and based on the time stamps computes the number of independent MBs in each time slot. Figure 3.5 shows how the number of independent/parallel MBs varies over time during the decoding of one frame for four frame resolutions: 576p (SD), 720p (HD), 1080p (FHD) and 2160p (QHD)

The figure shows what was remarked before: at the beginning and end of processing a frame there are very few MBs that can decoded in parallel. We refer to the phase until maximum parallelism is achieved as the *ramp-up phase* and the phase in which the parallelism decreases as the *ramp-down phase*. For 1080p, the maximum parallelism is 60, but the average parallelism is "only" around 30.

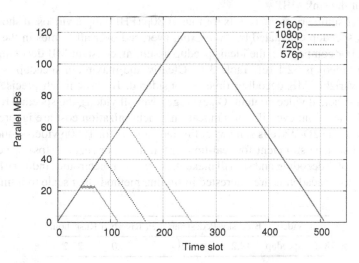

Fig. 3.5: Number of parallel MBs over time using the 2D-Wave approach for a single frame and four resolutions.

The analysis above is based on the wrong assumption that the MB decoding time is constant. In reality, however, the MB decoding time varies significantly, because although the same set of kernels are applied to each MB, the time each kernel takes depends on certain conditions of the image samples. This is the result of using adaptive kernels such as the deblocking filter and motion compensation. Specifically, the MB decoding time depends on certain input conditions such as the type and position of the MB, type of motion compensation interpolation (integer, fractional), type of deblocking filter, number of non-zero coefficients, etc. These values cannot be easily predicted or known in advance.

Why should we care about variable MB decoding times? Well, it affects how the algorithm should be implemented as well as its scalability. It affects the implementation because a static schedule where first the MBs on the first knight-jump-diagonal are processed in parallel, then the MBs on the second diagonal, etc. will not be very efficient because the time for each diagonal will be determined by the longest MB decoding time. We return to this issue in Chapter 4. It affects scalability because the depth of the DAG might have increased by a larger factor than the amount of work.

In order to analyze the scalability of the 2D-Wave if the MB decoding time can vary, we developed a high-level simulator called *frame_sim*. The input of *frame_sim* is a trace of MB decoding times (obtained by instrumenting the FFmpeg H.264 decoder [2]), and it calculates the minimum time required to process all MBs in a frame by calculating the Earliest Decoding Time (EDT) of each MB. The EDT of the source MB (again, $MB(0,0)$) is given by the decoding time of $MB(0,0)$. The EDT of all other MBs is given by the maximum EDT of all MBs on which it depends plus its own decoding time. The minimum time to process a frame is then given by the EDT of the sink MB.

Table 3.2 summarizes the results for four 1080p (FHD) input videos. It shows the maximum speedup calculated by *frame_sim*, averaged over all frames in the video sequence. For comparison, the ideal speedup assuming constant MB decoding time for 1080p videos is 32.1 (cf. Table 3.1). Clearly, the maximum speedup is lower when the variable MB decoding time is considered. By how much precisely depends on the actual video content. On average, over all videos, the speedup reduces by 33%. Conclusion: even if communication/synchronization cost are ignored, the scalability of the 2D-Wave is somewhat limited, at least for 1080p video sequences. It will probably be sufficient for real-time (25 frames per second (fps) or perhaps even a bit more) decoding and so will make the consumer electronics industry happy, but people from academia are interested in finding methods that scale to infinity.

Input Video	Blue_sky	Pedestrian_area	Riverbed	Rush_hour
Max. Speedup	19.2	21.9	24.0	22.2

Table 3.2: Maximum speedup calculated by *frame_sim* that takes variable MB decoding time into account, for four different 1080p input video sequences.

3.3.3.2 3D-Wave

The previous section has shown that the 2D-Wave can achieve reasonable, but not astounding scalability. It would not scale to a future many-core containing 100 cores or more. The natural academic question to ask is: Can we do better? (Hint: yes) In this section we present our novel (at least it was in January 2008 when we first published it) *3D-Wave* approach.

The 3D-Wave algorithm is based on the observation that, because usually there is no superfast motion between consecutive frames, motion vectors are typically short. Because of this, the decoder does not have to wait until a reference frame has been completely decoded before starting the decoding of the next frame. For each MB it "only" needs to be ensured that its reference area (the area in the reference frame that is used to predict the MB) has been decoded before. Consecutive frames can therefore be decoded partially in parallel. Not fully parallel, as in frame-level parallelism, but partially, meaning that the processing of the predicted frame has to stay slightly behind the processing of the reference frame. This is illustrated in Figure 3.6 for two sample frames.

In this figure, the dotted MBs are "in-flight", meaning that they are currently being processed. The dotted MBs in the second frame can be processed because their reference areas in the first frame, indicated by hashed blocks, have already been decoded. Within each frame, the 2D-Wave approach is employed, to satisfy the intra-frame dependencies. Hence we refer to this approach as the 3D-Wave, the third dimension being the temporal domain.

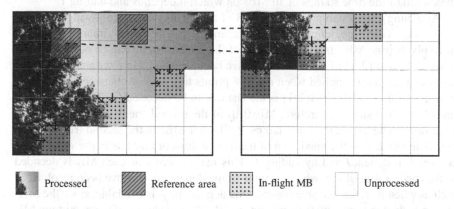

Processed   Reference area   In-flight MB   Unprocessed

Fig. 3.6: 3D-Wave strategy: frames can be decoded partially in parallel. Intra-frame dependencies are indicated by straight arrows and inter-frame dependencies by dashed arrows.

The 3D-Wave algorithm can be implemented either *statically* or *dynamically*. In the static approach, the worst case motion vector length is assumed for all MBs and a MB decoding task is scheduled no sooner than when it is certain that its reference area has been decoded. In the dynamic approach the actual motion vectors are considered and a MB decoding task is started as soon as its reference area has been decoded. Below we analyze the amount of parallelism both approaches exhibit.

Static 3D-Wave

In the static 3D-Wave the same maximum motion vector (MV) length is assumed for all MBs. Indeed, the H.264 standard defines a limit on the MV length. For 1080p resolutions and beyond, the limit is 512 pixels. We will exploit this static limit in Chapter 7 where we will present a highly efficient implementation of the static 3D-Wave.

To analyze the static 3D-Wave algorithm, we first assume that it takes constant, unit time to decode a MB, as we did for the 2D-Wave. In addition, the following conservative assumptions are made. First, B-frames can be used as reference frames. This represents the worst case for the 3D-Wave because it means that any frame can be used as a reference frame. Second, the reference frame is always the previous one. Again, this corresponds to the worst case because if the reference frame would be further in the past, there would be more slack between the processing of the reference frame and the frame it predicts. Third, only the first frame in the entire sequence is an I-frame.

Under these assumptions, the amount of MB-level parallelism exhibited by the static 3D-Wave can be calculated in a manner similar to as we did for the 2D-Wave. Each MB is assigned a time stamp, corresponding to the time slot in which it can be decoded. The time stamp of the first (the upper-left) MB in the first frame is 0, and the time stamps of all other MBs in the first frame are calculated by taking the maximum of the time stamps of all MBs on which it depends and adding 1.

Assigning a time stamp to the first MB of the second and all following frames is a bit tricky. Therefore, let us consider an example first. If the maximum MV length is 16 pixels (one MB), then the first MB of the second frame can be decoded when MBs $(1,2)$ and $(2,1)$ of the first frame have been decoded. Here we account for the interpolation that is applied when the MV points to a position between pixel coordinates. Of these MBs, $MB(2,1)$ is the last one to be decoded (in 2D-Wave order), namely in time slot 6. Therefore, $MB(0,0)$ of the second frame can be decoded in time slot 7. Thereafter, the time stamps of all other MBs in the second frame can be calculated by taking the maximum of the time stamps of all MBs *in the same frame* on which it depends and by adding 1. This makes sure that each MB is decoded in the earliest time slot in which its intra-frame dependencies have been resolved as well as when its reference area (inter-frame dependency) is available, since the MBs stay in lock-step sync with their reference MBs. As just shown, if the maximum MV length is 16, the delay between the time slot in which a MB can be decoded and the time slot in which the corresponding MB in its reference frame can be decoded is 7 MB decoding time units. Using similar reasoning, we find that for maximum MV lengths of 32, 64, and 128 pixels, the delay is 9, 15, and 27 MB decoding time units, respectively. In general, if the maximum MV length is n pixels ($\lceil n/16 \rceil$ MBs), the required delay is $3 \times \lceil n/16 \rceil + 3$ MB decoding time units.

Based on the analysis above, we wrote a program that, given the frame resolution and the MV range, computes the number of MBs that can be processed in each time slot. Figure 3.7 depicts the results for an FHD (1080p) sequence and for power-of-2 MV ranges from 16 to 512 pixels. 512 is the maximum vertical MV length allowed for HD (720p) and FHD (1080p) levels (4.0 to 5.1) of the H.264 standard [1]. It can be seen that the static 3D-Wave algorithm has a large amount of parallelism to offer, much larger than the 2D-Wave. For example, when the maximum MV is 64 pixels, the number of MBs that can be processed in parallel, and hence the number of cores that may be employed efficiently, is 624. For the same resolution, the 2D-Wave achieves maximum speedup of 32.1 (cf. Table 3.1). The figure also shows that at the beginning there is a ramp-up phase where fewer MBs can be processed in parallel. In a real video sequence with many thousands of frames, however, this phase will be negligible, and is visible here only because we show the first 750 MB decoding time slots. In addition, the figure shows that the longer the MVs, the less parallelism the 3D-Wave approach exhibits. This can be expected but we remark again that very long MVs are very unlikely due to inertia in video sequences. Finally, the astute reader will notice that for very large maximum MV lengths (512 or 256 pixels), the amount of MB-level parallelism exhibits a tiny ripple. This is because when MVs are very long, the ramp-down phase of the intra-frame, 2D-Wave parallelism cannot be completely concealed anymore.

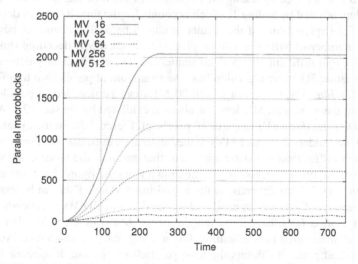

Fig. 3.7: Number of MBs that can be decoded in parallel using the static 3D-Wave approach in each MB decoding time slot for an FHD video sequence and several MV ranges.

Dynamic 3D-Wave

The static 3D-Wave approach has a large amount of parallelism to offer, probably (but this is to be verified) more than is required to achieve real-time. A developer working in industry will probably say, that's good enough, and he or she is probably right since time is money. Applied researchers, however, are driven by two fundamental questions: Can we do better? and What is needed to effectively implement it? This time we will provide no hints to build up the tension a little.

The dynamic 3D-Wave algorithm does not assume the same worst case MV for all MBs, but considers the actual MVs and reference areas of each MB. This implies that the amount of MB-level parallelism cannot be determined analytically anymore. Instead, a trace-driven approach is needed that reads in a trace of MB coordinates and their MVs, and based on these, calculates the earliest MB decoding time slot in which each MB can be processed. As before, we assume it takes unit time to decode a MB.

For this we have developed a high-level simulator called *3DWave_sim* which is similar to *frame_sim* that was used to analyze the 2D-Wave approach. *3DWave_sim* works as follows. As in the previous analysis, it assigns a time stamp to each MB corresponding to the earliest time slot in which the MB can be decoded. Furthermore, *3DWave_sim* processes all frames in decode order and not in display order, and the MBs within each frame in scan order (hence, as the MBs are encoded in the bitstream). This assures that when a time stamp is assigned to a MB, all MBs on which it can depend have already been assigned a time stamp. The time stamp of each MB *B* is calculated by taking the maximum of the time stamps of all MBs on which *B* depends and by adding 1, which accounts for the time it takes to decode *B*.

Figure 3.8 depicts some of the results produced by *3DWave_sim*. As before, it shows the number of MBs that can be processed in each MB decoding time slot, this time for four different 1080p video sequences. The results clearly demonstrate that the dynamic 3D-Wave algorithm has a *huge* amount of parallelism to offer. For example, for *Blue_sky*, up to roughly 7000 MBs can be decoded in parallel. For comparison, the maximum MB-level parallelism exhibited by the static 3D-Wave is around 2000 MBs (for a MV range of 16 pixels, cf. Figure 3.7). So the dynamic approach, that considers the actual MVs in the bitstreams, even has substantially more parallelism to offer than the static approach, that assumes the worst case MV for all MBs. Several other conclusions can be drawn from the figure. First, the amount of MB-level parallelism depends on the actual input video. This can be expected, since the amount of parallelism exhibited by the dynamic 3D-Wave depends on the amount of motion (and, hence, the lengths of the MVs) in the video. For example, the *Pedestrian_area* movie features fast moving objects, which result in many large MVs, and hence it exhibits the least parallelism. Second, it appears that the dynamic 3D-Wave suffers from a ramp-up and a ramp-down phase with in between a longer or shorter stable phase where the MB-level parallelism stays more or less constant. This is mainly because "only" 400 frames were simulated and because the entire sequences are hypothetically decoded so fast. For real video sequences with many thousands of frames and on real many-core systems containing at most a few hundreds cores, the stable middle phase will be much longer and the ramp-up and ramp-down phases will be negegible. Normally we think twice before making bold claims, but it seems more likely that Moore's law will come to an end before the parallelism of the dynamic 3D-Wave is exhausted, at least on a single processor. GPUs, however, are a different story.

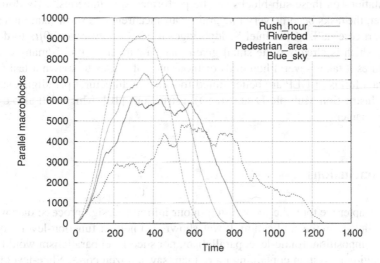

Fig. 3.8: Number of MBs that can be decoded in parallel using the dynamic 3D-Wave approach in each MB decoding time slot for several FHD video sequences.

One potential drawback of the 3D-Wave is that it requires a large number of frames in-flight, i.e., a large number of frames to be decoded simultaneously. When it comes to implementing the algorithm, this can be a disadvantage since the *working set* (the data referenced by the application during a certain time interval) may be larger than can be kept on chip, causing the average memory access time to increase significantly. The article on which this chapter is based [5] also considers the effect of limiting the number of frames that can be in-flight simultaneously, as well as the effect of using different motion estimation algorithms. We have not analyzed the effects of variable MB decoding times for the 3D-Wave, as we did for the 2D-Wave. It is reasonable to assume, however, that it will reduce the scalability of the 3D-Wave by as much as it reduced the scalability of the 2D-Wave (about 33%). Furthermore, the 3D-Wave exhibits so much parallelism, that a reduction of even 50% will not impair its scalability. Analysis is nice and very useful, but it should not be carried too far.

3.3.4 Other Data-level Decompositions

There are other data decompositions possible besides the ones described in this section. For example, it might be possible to parallelize H.264 decoding at the GOP-level, meaning that several Groups-of-Pictures (GOPs) are processed in parallel, but this will also exhibit very limited scalability. Another approach, to extract even more parallelism than the 3D-Wave has on display, is to parallelize at the sub-block-level, i.e., to exploit the fact that MBs are further divided into several sub-blocks and that computations on these sub-blocks can be performed simultaneously. By doing so, however, the tasks become very fine grain. Our meauremunts show than on a state-of-the-art core (a 3.3 GHz Intel Sandybridge) it takes on average $2.0\mu s$ to decode a MB, which is already quite small given that on almost all multi-/many-core architectures it takes several hundreds or thousands of cycles to spawn a task. Such fine-grain forms of DLP are better suited for other architectural paradigms such as Single-Instruction Multiple-Data (SIMD) instructions [10], which will also exploit in later chapters.

3.4 Conclusions

In this chapter we took the second step in our informal design process: discovering the parallelism in the application. It was shown that neither function-level (pipelining) decomposition, frame-level parallelism, nor slice-level parallelism would scale on a multi-core system containing more than, say, a dozen cores. Slice-level parallelism might scale, but using more than just a few slices significantly decreases the coding efficiency, which is a no-no, a curse in the video coding domain.

MB-level parallelism is more promising, and therefore we analyzed it in more detail analytically as well as using high-level simulation. Such analyses are more cost-efficient than immediately implementing the algorithm only to discover that the amount of parallelism is insufficient to reach the targeted performance. Two algorithms based on MB-level parallelism were presented. The first one, called the 2D-Wave, exploits that the MBs on a knight-jump-diagonal inside a frame can be processed in parallel. Under the (wrong) assumption that the MB decoding time is constant, the analytical results show that the 2D-Wave can achieve a maximum speedup of 32.1. When the variable MB decoding time is taken into account, the high-level simulation results show that the maximum speedup decreases to about 22.0 on average, which is about a third less. The 3D-Wave algorithm exploits the same intra-frame MB-level parallelism as the 2D-Wave does, but also exploits MB-level between frames (the scientific sounding inter-frame MB-level parallelism). The simulation results show that the 3D-Wave exhibits huge amounts of parallelism. In one case more than 9000 MBs could be processed in parallel.

Discovering the parallelism is necessary but not enough, however. The application developer also has to find ways to efficiently implement the algorithms, i.e., to efficiently map the algorithm onto the architecture. Now issues such as communication/synchronization overhead and load balancing start to play major roles. Furthermore, while implementing the algorithm, the application developer often has to take the peculiarities of the targeted architecture into account. We will illustrate this step for the 2D-Wave in the next chapter and for the 3D-Wave in Chapter 5.

References

1. Advanced Video Coding for Generic Audiovisual Services, Recommendation H.264 and ISO/IEC 14 496-10 (MPEG-4) AVC, International Telecommunication Union-Telecommun. (ITU-T) and International Standards Organization/International Electrotechnical Commission (ISO/IEC) JTC 1 (2003)
2. FFmpeg (2011). http://ffmpeg.org
3. Flierl, M., Girod, B.: Generalized B Pictures and the Draft H. 264/AVC Video-Compression Standard. IEEE Transactions on Circuits and Systems for Video Technology **13**(7), 587–597 (2003)
4. Jacobs, T., Chouliaras, V., Mulvaney, D.: Thread-Parallel MPEG-2, MPEG-4 and H.264 Video Encoders for SoC Multi-Processor Architectures. IEEE Transactions on Consumer Electronics **52**(1), 269–275 (Feb. 2006)
5. Meenderinck, C., Azevedo, A., Alvarez, M., Juurlink, B., Ramirez, A.: Parallel Scalability of Video Decoders. Journal of Signal Processing Systems **57**, 173–194 (2009)
6. Roitzsch, M.: Slice-balancing H.264 Video Encoding for Improved Scalability of Multicore Decoding. In: Proceedings of the 7th ACM & IEEE International Conference on Embedded Software, pp. 269–278 (2007)
7. Sze, V., Chandrakasan, A.P.: A High Throughput CABAC Algorithm using Syntax Element Partitioning. In: Proceedings of the 16th IEEE International Conference on Image processing, pp. 773–776 (2009)
8. van der Tol, E.B., Jaspers, E.G.T., Gelderblom, R.H.: Mapping of H.264 Decoding on a Multiprocessor Architecture. In: Proceedings of SPIE (2003)
9. Wiegand, T., Sullivan, G.J., Bjontegaard, G., Luthra, A.: Overview of the H.264/AVC Video Coding Standard. IEEE Transactions on Circuits and Systems for Video Technology **13**(7), 560–576 (2003)
10. Zhou, X., Li, E.Q., Chen, Y.K.: Implementation of H.264 Decoder on General-Purpose Processors with Media Instructions. In: Proceedings of SPIE Conference on Image and Video Communications and Processing (2003)

Chapter 4
Exploiting Parallelism: the 2D-Wave

Abstract In the previous chapter we have analyzed various parallelization approaches for H.264 decoding and concluded that in order to scale to a large number of cores, macroblock-level parallelism needs to be exploited. The next question is how to efficiently exploit this parallelism. In other words, how to map this parallelism onto a multi-/many-core architecture. To answer this question, in this chapter we present two implementations of the 2D-Wave approach. The first implementation maintains a centralized pool of macroblocks that are ready to be decoded and cores retrieve tasks from this *Task Pool*. In the second approach, called Ring-Line, full lines of macroblocks are statically assigned to cores and the cores synchronize and communicate point-to-point. Both approaches have been implemented and are evaluated on a dual-chip Cell BE system with 18 cores in total.

4.1 Introduction

In Chapter 3 we have described the 2D-Wave algorithm and evaluated its scalability using an analytical approach, where we assumed that it takes constant time to decode a macroblock (MB) and that there is no communication and synchronization overhead. In reality, however, the MB decoding time varies significantly and communication and synchronization overhead needs to be minimized in order to obtain an efficient implementation. Thus the next question, which is addressed in this chapter, is how to efficiently implement the 2D-Wave algorithm.

Two important issues need to be considered when developing parallel applications: *load balancing* and *communication/synchronization overhead*. The load needs to be distributed equally over the cores so that cores are not idle waiting for results from other cores. For example, a static scheduling where first the first diagonal of parallel MBs is processed, then the second diagonal, and so on, cannot lead to an efficient implementation because the time to process a diagonal will be determined by the longest MB decoding task. Furthermore, the communication/synchronization overhead needs to be minimized, since when it is significant, the overall speedup will be limited. One technique to reduce communication overhead is by overlapping communication with computation.

This chapter presents two implementations of the 2D-Wave algorithm on a system consisting of two Cell BE processors. The first implementation, referred to as the Task Pool (TP) [2], is based on the master-slave programming paradigm. Slaves request work (in this case corresponding to MBs) from a master, which keeps track of the dependencies between the MBs. This implementation can achieve (in theory) perfect load balancing, since MBs are submitted to the task pool as soon as they can be decoded. In the second implementation [4], referred to as Ring-Line (RL), the cores process entire lines of MBs rather than single MBs. This approach does not require a centralized master to keep track of dependencies. Furthermore, its static mapping of MBs to cores allows overlapping communication with computation, since it is known a priori which core will decode which MB. A potential disadvantage of the RL implementation is, however, that it might cause load imbalance if MB lines have different processing times. These trade-offs are evaluated on an 18-core Cell BE system.

This chapter is organized as follows. In Section 4.2 a brief overview of the Cell architecture is presented. The TP implementation is described in Section 4.3 while Section 4.4 presents the RL approach. Experimental results are presented in Section 4.5, and conclusions are drawn in Section 4.6.

4.2 Cell Architecture Overview

The Cell Broadband Engine [7] (Cell BE) is a heterogeneous multi-core consisting of one PowerPC Element (PPE) and eight Synergistic Processing Elements (SPEs). The PPE is a dual-threaded general purpose PowerPC core with a 512 kB L2 cache. Its envisioned purpose is to act as the control/OS processor, while the eight SPEs should provide the computational power. Figure 4.1 shows a schematic overview of the Cell processor. The processing elements, memory controller, and external bus are connected to an Element Interconnect Bus (EIB). The EIB is a bi-directional ring interconnect with a peak bandwidth of 204.8 GB/s [3]. The XDR memory can deliver a sustained bandwidth of 25.6 GB/s.

Two features make the Cell processor an innovative design. First, it is a *functionally heterogeneous multi-core*, meaning that the cores are optimized for different types of code. In the Cell, the PPE is optimized for control/OS code, while the SPEs are targeted at throughput computing kernels. Second, it has a *scalable memory hierarchy*. In conventional homogeneous multi-core processors, each core has several levels of cache. The caches reduce the average latency and bandwidth requirements to the external (off-chip) memory. With multiple cores there are multiple caches and cache coherence is required to make sure that the data in the private cache of each core uses is up-to-date. The complexity of cache coherence increases with the number of cores, however. In the Cell architecture, the SPEs do not feature a cache but rely on a software-managed *local store* and a *Direct Memory Access* (DMA)

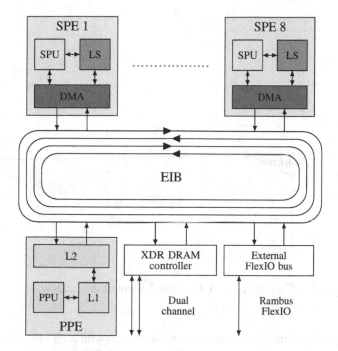

Fig. 4.1: Schematic view of the Cell Broadband Engine architecture.

unit instead to access the memory. Each SPE has a local store of size 256 kB and can only process data that is present in its local store. The programmer is responsible for transferring data from/to main memory using explicit DMA operations. The programming style resembles a *shopping list*. Instead of fetching every data block separately at the time it is needed (as is more or less done in cache-based systems), all data required by a task is brought in at once and before the execution of the task. Moreover, loading the data of the next task should be done concurrently with the execution of another task in order to hide the memory latency. This technique is referred to as *double buffering*.

4.3 Task Pool Implementation

In the Task Pool approach, a centralized master is used to dynamically distribute the MB decoding tasks over the processing elements (PEs). A high level view of the implementation is shown in Figure 4.2. The master task M in the figure is executed on the PPE and the SPEs are denoted by P_i. The SPEs are the slaves who inform the master when they are done executing a task and request a new task. The master keeps track of the dependencies between MBs using a dependency table and inserts MB decoding tasks that are ready to be executed into a task queue.

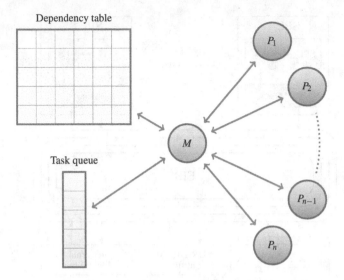

Fig. 4.2: Centralized Task Pool implementation of 2D-Wave.

The dependency table associates a dependency count with each MB in the frame. The value of the dependency count associated with MB *B* is equal to the number of unprocessed MBs on which *B* depends. Whenever an SPE has processed a MB *B*, it informs the master who then decrements the dependency counts associated with the right and down-left neighbors of *B*. A MB is ready to be decoded when its dependency count drops to zero. Ready MBs are scheduled first-in-first-out using the task queue. Figure 4.3 depicts the dependencies between MBs and the initial values in the dependency table. The initial values correspond to the number of arrows pointing to each MB. Note that Figure 3.3 in Section 3.3.3.1 on Page 23 shows that a MB can depend on up to four MBs. The dependencies from a MB to its down and down-right neighbors, however, are covered by the dependencies from a MB to its right and down-left neighbors.

Listing 4.1 depicts pseudo-code for initializing the dependency counters, where *WIDTH* is the width of each frame in MBs and *HEIGHT* is the height of each frame in MBs.

The dependency table and task queue are only accessed by the PPE, thereby providing synchronized access. It is therefore not necessary to atomically decrement the dependency counts, as is required in the shared memory implementation described by Hoogerbrugge and Terechko [6]. To issue work to an SPE, the PPE sends the MB coordinates to be decoded in a 32-bit mailbox message to the SPE. On completion, the SPE reports back to the PPE also with a mailbox message. Experiments, where we also considered mutual exclusion variables (mutexes) and atomic instructions, have shown that this is the fastest master-slave implementation. This scheme requires, however, the PPE to continuously loop over the mailbox statuses.

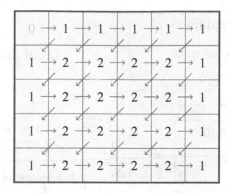

Fig. 4.3: MB dependencies and initial values in the dependency table. After a MB is processed, the master decrements the dependency count associated with the MBs that are pointed to by the arrow(s).

```
init_dependency_matrix()
{
    dep_count[0][0] = 0;
    for (j=1; j<WIDTH; j++)
        dep_count[0][j] = 1;
    for (i=1; i<HEIGHT; i++)
        dep_count[i][0] = dep_count[i][WIDTH-1] = 1;
    for (i=1; i<HEIGHT; i++)
        for (j=1; j<WIDTH-1; j++)
            dep_count[i][j] = 2;
}
```

Listing 4.1: Initialization of the dependency table.

In shared memory systems, the cores can access the entire frames located in the shared memory. The SPEs cannot access the entire frames, however, since the frames are too large to fit in the local store. Furthermore, to conserve memory bandwidth, only the frame parts that are needed to perform the task should be brought into the local store. A block of 48×20 pixels is allocated in the local store to serve as the working buffer for the MB decoding kernels. This block size is used to store the intra-prediction data shown in Figure 3.2 on Page 22. Although the actually needed data is only 28 pixels wide, due to DMA *alignment restrictions*, a larger buffer is required. On the Cell, each DMA must be 16 byte aligned and the DMA size must be a multiple of 16 bytes. Therefore, the buffer width corresponds to the width of three MBs. For similar reasons the size of the reference data buffer is larger than the actual data.

Figure 4.4 illustrates the DMA transfers between external memory and the local store. Between the DMA transfers the MB decoding kernels are applied. Intra-prediction is applied after (2) and motion compensation after (3), followed by inverse quantization and the IDCT. After (4) the deblocking filter is applied. In the

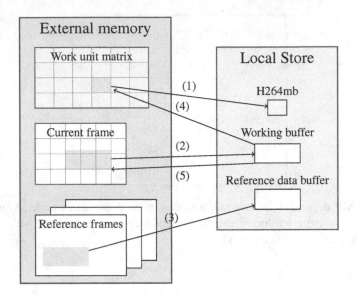

Fig. 4.4: DMA transfers needed to decode a single MB. (1) After receiving the MB coordinates, the corresponding H264mb structure is transferred to the local store. (2) This is used to fill the working buffer with the correct intra data. (3) If motion compensation is required, the relevant part of the reference data is retrieved for each partition. (4) Before applying the deblocking filter, the unfiltered borders are stored in the surrounding H264mb structures. (5) After performing all the MB kernels the working buffer is written back.

task pool implementation of the 2D-Wave the DMA transfers cannot be overlapped with computation, because the input data can only be transferred after the coordinates of the MB are known and the output data has to written back before notifying the master PPE of completion. Moreover, DMA transfers (1) and (3) also have to be executed consecutively, as the required motion vectors are only available after DMA transfer (1).

Listing 4.2 depicts pseudo-code for slave SPEs. The slave SPE code is relatively simple. It consists of an infinite loop in which each SPE first receives the coordinates of a MB that is ready to be decoded. In the actual implementation there is no infinite loop but the SPEs are informed by the master when the entire video sequence has been decoded. After receiving the MB coordinates, the SPE fetches the MB data (represented by an H264mb data structure) and its intra data from the memory into its local store. The MB data also contains the motion vector information, so only after the MB data has been fetched, the reference area can be fetched, which is why a blocking DMA transfer is needed. Thereafter, the MB is actually decoded, the decoded MB is written back to memory so that other MB decoding tasks can fetch their intra data from there, and the MB coordinates are sent back to the master to inform the master that it has been processed.

```
while (!Finished){
    (x,y) = wait_for_next_MB();
    H264mb = fetch_MB_data(x,y);
    working_buf = fetch_intra_data(x,y);
    ref_data_buf = fetch_reference_area(x,y);
    decode_MB(x,y);
    write_MB(x,y);
    notify_master(x,y);
}
```

Listing 4.2: Pseudo-code for the slave SPEs.

The pseudo-code for the master PPE depicted in Listing 4.3 is slightly more complex. It consists of a loop that is repeated for each frame. In each loop iteration first the matrix of dependency counters is initialized, using the pseudo-code given in Listing 4.1. The master keeps track of the number of MBs that have been decoded using a counter decoded_mbs, which is initialized to 0. To be able to assign work to idle SPEs, the master uses a queue of core identifiers (IDs) of cores that are ready to receive a new MB decoding task. To lead off, the coordinates of the MB with no input dependencies (MB(0,0)) are enqueued in this ready queue. Thereafter, the PPE enters a loop which is executed as long as the number of decoded MBs is less than the number of MBs in a frame. In each iteration of this loop, the master PPE first assigns ready MB decoding tasks to idle SPE slaves, as long as there are MBs ready to be decoded and as long as there are SPEs waiting to receive a new task. Then the mailbox of each SPE is scanned. If there is a message in the mailbox of SPE i corresponding to the coordinates of a MB (x,y) that has been decoded by SPE i, the PPE increments the number of decoded MBs and adds SPE i to the queue of ready cores. Thereafter, the master decrements the dependency counters of the right and upper-left neighbors of MB (x,y) (provided they exist). If the dependency counters become 0, the corresponding MBs are added to the queue of MBs that are ready to be decoded.

4.4 Ring-Line Implementation

In the Ring-Line implementation, a more static approach is used to leverage MB-level parallelism. Instead of individual MBs, the SPEs process entire lines of MBs. Figure 4.5 illustrates the mapping of MBs to SPEs for the case of 3 SPEs. In this example, SPE P_1 processes the first and last MB line, SPE P_2 the second MB line, and SPE P_3 the third MB line. The Ring-Line implementation increases data locality (because horizontally adjacent MBs are processed by the same SPE), allows to overlap communication with computation (because it is known a priori which SPE will need the produced data), and reduces synchronization overhead (because no central master is needed to keep track of the decoding status of each MB).

```
for (each frame)
{
    init_dependency_matrix();
    decoded_mbs = 0;
    enqueue(ready_q, (0,0));

    for (i=0; i<NSPEs; i++)
        enqueue(core_q, i);
    while (decoded_mbs < HEIGHT*WIDTH)
    {
        while (!(is_empty(ready_q) || is_empty(core_q)))
        {
            (x,y) = dequeue(ready_q);
            i = dequeue(core(q));
            send_MB_coordinates_to_spe(x,y,i)
        }
        for (i=0; i<NSPEs; i++)
        {
            if (mailbox[i]){
                decoded_mbs++;
                enqueue(core_q, i);
                (x,y) = read_mailbox(i);
                if (y<WIDTH-1)
                    if (--dep_count[x][y+1]==0)
                        enqueue(ready_q, (x,y+1));
                    if (x<HEIGHT-1 && y>0)
                        if (--dep_count[x+1][y-1]==0)
                            enqueue(ready_q, (x+1,y-1));
            }
        }
    }
}
```

Listing 4.3: Pseudo-code for the master PPE.

Figure 4.5 shows that only the MBs on the same line and the next line depend on the MBs processed by a certain SPE. This static mapping of MBs to SPEs allows the right column of a MB B_1 that is needed to decode the MB B_2 to the right of MB B_1 to stay in the local store of the SPE that processes both MBs. Furthermore, the bottom rows of B_1 can be sent via the on-chip interconnect to the SPE that processes the next MB line (cf. Figure 4.1 on Page 37). The SPEs are, therefore, logically organized in a ring network, hence the name Ring-Line, which is abbreviated as RL. Furthermore, synchronization is no longer centralized but distributed. SPEs signal the completion of a MB by sending non-blocking, one-way control signals to the next SPE, together with the intra data, rather than to a central master.

Because it is known that the next core in the ring will require the produced intra data, this data can be *pre-sent* before it is needed. To support this, local buffers are required. In the implementation we use a buffer with a width of an entire line and a height of 20 pixels. This line buffer is used as the working buffer as well as for

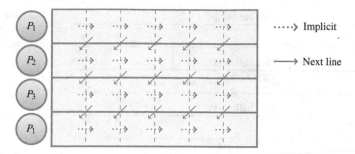

Fig. 4.5: Dependencies in the Ring-Line algorithm. Dependencies to blocks on the same line are implicit, i.e., they are satisfied by the order in which each SPE processes the blocks.

the intra data. By directly writing the intra data to the right place in the line buffer, additional copy steps are avoided. In the RL implementation all intra data is kept on-chip.

Once a MB has been decoded, it needs to be written back to the frame buffer stored in external memory. Because of DMA size and alignment restrictions, two adjacent MBs are written back together. Additionally, the write back step is delayed by one MB, enabling the deblocking filter to modify the data before writing it back to the frame buffer. This results in writing each pixel in the frame only once, while the TP implementation writes it 3 to 6 times. The orchestration of the communication in the RL implementation is illustrated in Figure 4.6. In the figure c (short for current) denotes the MB that is currently being decoded. The bottom rows of two previously decoded MBs, which is part of the intra data required by the next line, are pre-sent to the line buffer of the next SPE. This transfer is issued together with the write back transfer of the two previously decoded MBs to the frame buffer. Additionally the reference data (r) and the work unit (w) of the next MBs to be processed are *prefetched*, meaning that they are fetched from the off-chip memory before they are actually needed in order to hide the memory latency. Hece all DMA transfers are performed concurrently with the processing of the current MB (c).

Figure 4.7 shows how the DMA transfers and the processing steps are scheduled. In, e.g., the fifth time slot, the work unit of MB 5 is prefetched into the local store, the reference data of MB 4 is also prefetched, MB 3 is being processed, and MBs 1 and 2 are written back together to the frame buffer in external memory. Each DMA transfer (indicated by a non-rounded rectangle) is *non-blocking* and its completion is checked before issuing the same step in the next time slot. This scheme completely hides the DMA latency as long as the DMA latency is shorter than the MB processing time.

High-level pseudo-code for the RL implementation is depicted in Listing 4.4, where for brevity and comprehensibility, the start-up code and the finish-up code have been omitted. As illustrated in Figure 4.7, the start-up pseudo-code corresponds to the first four phases of the schedule of the DMA transfers / processing

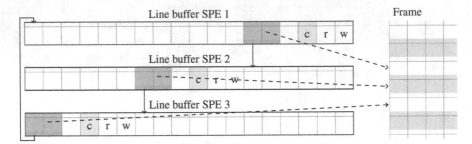

Fig. 4.6: Orchestration of communication in the RL implementation. All DMA tranfers are performed concurrently to the decoding of the current MB (c).

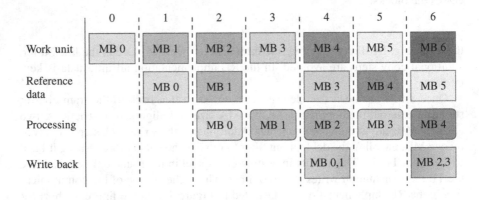

Fig. 4.7: Scheduling of DMA transfers and processing steps. The DMA transfers indicated by non-rounded rectangles are non-blocking. Their completion is checked in the next slot.

phases, where some DMA transfers or the processing step are empty. Similarly, there will be finish-up code at the end of processing each MB line. Several other simplifications have been made as well.

The pseudo-code listed in Listing 4.4 is repeated for each frame in the video sequence. For each frame, the SPE with identifier `spe_id` processes the MB lines `spe_id`, `spe_id+NSPEs`, `spe_id+2*NSPEs`, and so on. In other words, the MB lines are distributed in a round-robin fashion over the SPEs. For each MB line some start-up code needs to be executed, which, as described above, is omitted for brevity and comprehensibility. Let x be the MB line that is currently being processed. Each SPE then loops over all MBs in the current line, starting with fetching $MB(x,4)$, since this corresponds to the first phase without empty DMA transfer / processing slots. In each iteration y of this loop, a number of non-blocking DMA transfers are started, fetching to work unit for $MB(x,y+2)$, the reference area for $MB(x,y+1)$ (which, as explained before, can only be fetched after the work unit

```
for (each frame)
{
    for (x=spe_id; x<HEIGHT; x+=NSPEs)
    {
        // start-up code (omitted)
        for (y=2; y<WIDTH; y++)
        {
            dma_get( work_unit(x,y+2) );
            wait_on( work_unit(x,y+1) );
            dma_get( ref_data (x,y+1) );
            wait_on( ref_data (x,y  ) );
            if (y%2==0)
            {
                wait_for_prev_spe ( intra_data((x,y),(x,y+1)) );
            }

            decode_MB(x,y);

            if (y%2==0)
            {
                send_to_next_spe( intra_data((x,y-2),(x,y-1)) );
                // write back
                dma_put( decoded_data((x,y-2),(x,y-1)) );
            }
        }
        // finish-up code (omitted)
    }
    dma_barrier(); // wait until all dma transfers have completed
}
```

Listing 4.4: Pseudo-code for SPE in the RL approach.

has been fetched), and, if y is even, the combined write back step of $MB(x, y - 2)$ and $MB(x, y - 1)$. The wait_on and wait_for_prev_spe are used to check the completion of non-blocking DMA transfers issued in previous iterations.

The DMA command dma_get is used to start a DMA transfer from main memory to the local store, while dma_put is used to start a DMA transfer from the local store to main memory. Thereafter, the SPE checks if it has received a synchronization message and the intra data from its predecessor in the logical ring network. This data is not communicated via the memory but is sent over the on-chip interconnect. In the next step, $MB(x, y)$ is decoded, and only for even MBs a synchronization message and the intra data is sent to the next SPE in the logical ring network. Before starting processing the next frame, a dma_barrier is executed. This ensures that the current frame (which could be a reference frame for the next frame) has been completely decoded before the decoding of the next frame is started. Provided the MB decoding time is longer than the DMA latency, the communication time should be hidden completely behind the computation time.

4.5 Experimental Evaluation

The efficiency of the parallel implementations was evaluated on a Cell Blade consisting of two Cell processors connected via a FLEXIO bus, which has a peak bandwidth of 37.6 GB/s. The second Cell processor shares the memory controller of the first via the FLEXIO bus. The amount of XDR external memory is 1 GB with a rated bandwidth of 25.6 GB/s. The operating system is Fedora Core 7 with kernel version 2.6.22.

The FFmpeg [5] audio/video decoder is used as a baseline for our parallel implementations. The FFmpeg build is configured to use all optimizations, including the AltiVec PowerPc extensions. The SPE compiler optimization setting is -O2. The PPE and SPE hardware counters, which have a resolution of 14.8 MHz and negligible call time, are used to measure execution times.

The experiments are performed with the Full High Definition (FHD) sequences from the HDVideoBench [1]. These sequences are encoded with X264 [8] and conform the High profile level 4.0 H.264 standard using CABAC as the entropy decoding scheme. More specifically, the streams are encoded using two B-frames between I- and P-frames with weighted prediction. The motion vector range is set to 24 pixels.

To implement the 2D-Wave parallel approaches on the Cell processor, the SPEs have to apply the MB reconstruction and filtering kernels. The task of the PPE is to perform the entropy decoding and act as the controller. In the original code the MBs are processed in scan line order. This includes the entropy decoding, which extracts the parameters for one MB at a time. Therefore, for both implementations the entropy decoding has to be decoupled. The output of the entropy decoding is stored in a work unit matrix of H264mb structures for each MB.

Also the MB kernels need to be ported to SPE code. The motion compensation and IDCT kernel are ported to use the SPE SIMD engine using the FFmpeg Altivec code as a base. The deblocking filter and intra-prediction use scalar code.

In this chapter we only present results for the parallelization of the MB decoding stage. We assume that either an entropy decoding accelerator exists or that the entropy decoding has been parallelized at a different level. In Chapter 6 this problem is analyzed in detail.

4.5.1 Performance and Scalability

Figure 4.8 depicts the performance in frames per second (fps) of both implementations as a function of the number of SPEs for several HDVideoBench sequences. It can be seen that the RL implementation achieves much higher performance than the TP implementation. For example, on 16 SPEs, averaged over the video sequences, RL achieves a performance of 139.6 fps while the TP implementation attains only 76.6 fps. This performance difference can be partly explained by the base performance, i.e., the performance on a single core, which is higher for the RL implementation, as can also be seen in Figure 4.8. The other reason is the scalability of both implementations, as will be discussed next.

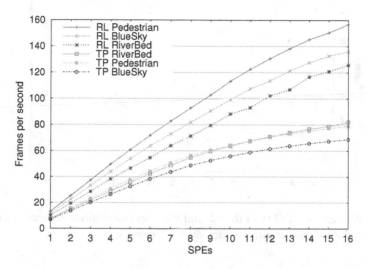

Fig. 4.8: Performance in frames per second of the HDVideoBench FHD BlueSky, Pedestrian and RiverBed sequences.

Figure 4.9 depicts the scalability of both implementations, i.e., the speedup over itself running on a single core. In order to improve legibility, in all cases the speedup is averaged over the video sequences. Included in the figure are also ideal speedups assuming no communication and synchronization overhead (the curves labeled "TP theoretical" and "RL theoretical"). These results have been obtained using a trace-driven simulator that simulates both implementations using a trace of MB decoding times, similar to the simulator described in Section 3.3.3.1.

Several conclusions can be drawn from Figure 4.9. First and foremost, the actual scalability of the RL implementation at 16 SPEs is about 15-20% higher than the actual scalability of the TP implementation. The theoretical scalability obtained using the simulator, on the other hand, indicates that TP is more scalable than RL, because TP should achieve perfect load balancing since all MBs that are ready to be decoded are kept in a centralized task pool. In other words, TP should incur fewer *dependency stalls* than the RL implementation, which is a term we use for the situation that a MB (or, in general, a task) cannot be processed yet because the data on which it depends is not available yet. In reality, however, the TP implementation cannot achieve the ideal scalability due to communication and synchronization overhead. Communication overhead is incurred because TP requires blocking DMA transfers. The DMA latency can therefore not be hidden. Furthermore, because it is unknown a priori which SPE is going to decode which MB, results need to be communicated via shared memory, causing shared memory contention. The TP implementation also incurs significant synchronization overhead due to the central

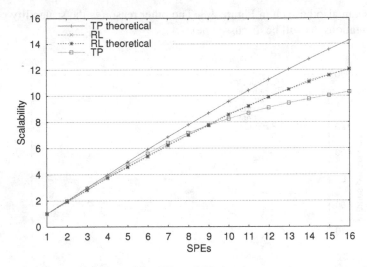

Fig. 4.9: Average scalability of the TP and RL implementations compared to theoretical speedups obtained using a trace-driven simulator.

master. When an SPE has processed a MB, it needs to request a new MB decoding task from the master. At that time, however, the master might be busy attending to another slave SPE and therefore the SPE has to wait. Such a central master will inevitably become a bottleneck, especially at higher core counts. The actual scalability of the RL implementation, on the other hand, almost exactly matches the theoretical scalability, Because the DMA latency can be hidden, the intra data can be kept on chip, and the SPEs synchronize point-to-point (in other words, this implementation features a distributed control mechanism), it incurs almost no communication and synchronization overhead. The disadvantage of the RL implementation is that it suffers more from load imbalance (dependency stalls) than the TP implementation, as is confirmed by the theoretical scalability results. On the Cell platform as well as many other platforms, however, it is in this case more important to minimize communication/synchronization overhead than it is to balance the load on each core.

4.5.2 Profiling Analysis

We further investigate the performance advantage of RL over TP by profiling the execution of the BlueSky sequence, which uses a lot of motion compensation.

Figures 4.10(a) and 4.10(b) show the profiling results for the TP and RL implementations, respectively. The figures break down the average MB decoding time into processing time, DMA cost, time needed for synchronization and dependency stalls, and time lost because there are fewer MB decoding tasks than SPEs (during the ramp-up and ramp-down phases of the 2D-Wave as described in Section 3.3.3.1

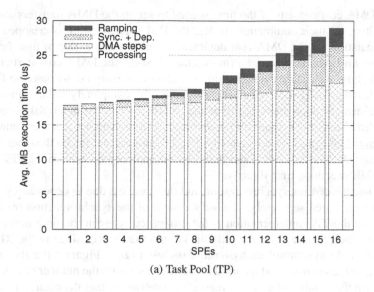

(a) Task Pool (TP)

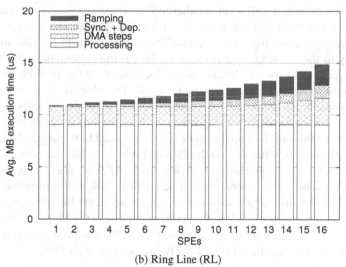

(b) Ring Line (RL)

Fig. 4.10: Breakdown of the average MB execution time for the TP and RL implementations using the BlueSky sequence.

on Page 25). In both figures the MB processing time stays constant, as can be expected. The relative time lost due to the ramping phases is also as expected and similar for both implementations. The DMA cost and the time needed for synchronization and dependency stalls, however, are much higher for TP and, furthermore, increase with the number of SPEs.

The DMA cost consists of the time needed to set up the DMA transfers and the time waiting for their completion. In RL, the DMA transfers are overlapped with the computation, so the DMA cost depicted in Figure 4.10b is due to time needed for setting up DMA transfers. For more than 13 SPEs, the DMA cost starts to increase slightly, which indicates that due to *network/memory contention*, the DMA latency has increased to a point where it cannot be completely hidden anymore. Network/memory contention describes the situation where so much data needs to be transferred over the network or read/written from/to memory, that the communication cannot take place simultaneously. The TP implementation, on the other hand, cannot hide the DMA latency, and therefore the memory contention increases the average MB execution time directly.

The time needed for synchronization and the time lost due to dependency stalls cannot be measured separately. A good estimate of the synchronization overhead incurred by the TP implementation is the difference between the time needed for synchronization and dependency stalls in TP and that in RL, since in the RL implementation the synchronization overhead is close to zero. Figure 4.10a shows that the synchronization overhead incurred by TP increases with the number of cores because when the number of cores increases, the probability that the central master is busy attending to a slave SPE when another SPE requests a new task also increases. It can also be observed that when moving from 8 to 9 SPEs, the synchronization overhead increases more than it did before. This is partly due to the additional off-chip latency to the second Cell processor. In the RL implementation, on the other hand, the synchronization overhead does not increase at all because the SPEs synchronize point-to-point in a logical ring network. The observable increase of the synchronization/dependency stall time is solely due to dependency stalls.

Finally, to put the results into perspective, the base performance (performance on a single SPE) of the parallel implementations is compared to the original sequential FFmpeg implementation running on the PPE in Table 4.1. The results show that the sequential FFmpeg implementation *running on the PPE* is faster on average than the parallel implementations running on a single SPE. This is not surprising as any parallel implementation incurs some overhead that sequential implementations do not incur. The performance of RL is, however, very close to the FFmpeg/PPE performance and even exceeds it for the BlueSky sequence. This is quite good considering that in terms of chip area, the PPE is about twice as large as an SPE.

Sequence	Sequential	Task Pool	Ring-Line
BlueSky	11.5 fps	6.8 fps	11.7 fps
Pedestrian	14.8 fps	8.0 fps	13.1 fps
RiverBed	11.5 fps	7.5 fps	10.1 fps

Table 4.1: Single-core performance in frames per second (fps) of the original sequential FFmpeg and the parallel implementations. The original sequential FFmpeg results are obtained on the PPE. For the TP and RL a single SPE is used.

4.6 Conclusions

In this chapter we have presented and evaluated two implementations of the 2D-Wave algorithm. The first is based on a centralized Task Pool (TP) and dynamically distributes MB decoding task over the slave cores which perform the tasks and request a new task from the master when they are done. This implementation should achieve perfect load balancing since all MBs that are ready to be decoded are available in a centralized place. The second implementation statically assigns whole lines of MBs to each core. This provides several advantages. First, synchronization overhead is minimized because a core only needs to synchronize with the core that processes the next MB line rather than with a central master. Second, communication overhead is reduced because it is know a priori which core will require the data produced by a certain task. This allows the data to be communicated on-chip, rather than via shared memory. Furthermore, the RL implementation allows to hide communication cost by overlapping communication with computation. A disadvantage of the RL implementation is, however, is that it might incur more load imbalance than the TP implementation, because MBs are statically assigned rather that dynamically distributed. Our results collected on an 18-core dual Cell BE system show, however, that in this case it is more important to minimize communication/synchronization overhead than it is to balance the load. Indeed, on 16 SPEs the RL implementation achieves a performance of 139.6 frames per second (fps), while the TP implementation achieves only 76.6 fps.

This chapter has shown that in order to develop highly efficient parallel applications, it is not sufficient to discover the parallelism, as we did in the previous chapter. The application developer also needs to think about how to efficiently implement the parallelism exhibited by the application. He or she needs to worry about issues such as reducing synchronization overhead, minimizing communication cost by optimizing data locality, load balancing, latency hiding, overlapping communication with computation, etc. These issues make parallel programming much more complex than sequential programming. Better hardware and programming/tool support is needed in the future to relieve the parallel programmer from these burdens, and the parallel-appication-design-process we propose in this book is a small step in that direction.

The implementations discussed in this chapter achieve (higher than) real-time performance for FHD (1080p) video sequences. Higher performance, however, might be required for more complex video coding standards (such as the approaching HEVC standard), higher resolutions, and/or more complex applications such as 3D TV. To be able to scale beyond the 16 cores considered in this chapter, a parallelization approach that is able to extract more parallelism from the application than the 2D-Wave is needed. This is the step we take in the next chapter, where we discuss how the 3D-Wave can be efficiently implemented on a shared-memory system.

References

1. Alvarez, M., Salami, E., Ramirez, A., Valero, M.: HD-VideoBench: A Benchmark for Evaluating High Definition Digital Video Applications. In: Proceedings IEEE International Symposium on Workload Characterization (2007). http://personals.ac.upc.edu/alvarez/hdvideobench/index.html
2. Alvarez-Mesa, M., Ramirez, A., Azevedo, A., Meenderinck, C., Juurlink, B., Valero, M.: Scalability of Macroblock-level Parallelism for H.264 Decoding. In: Proceedings International Conference on Parallel and Distributed Systems (2009)
3. Chen, T., Raghavan, R., Dale, J., Iwata, E.: Cell Broadband Engine Architecture and its First Implementation: a Performance View. IBM Journal of Research and Development **51**(5) (2007)
4. Chi, C.C., Juurlink, B., Meenderinck, C.: Evaluation of Parallel H.264 Decoding Strategies for the Cell Broadband Engine. In: Proceedings of the 24th ACM International Conference on Supercomputing (2010)
5. The FFmpeg Libavcodec. http://ffmpeg.org
6. Hoogerbrugge, J., Terechko, A.: A Multithreaded Multicore System for Embedded Media Processing. Transactions on High-Performance Embedded Architectures and Compilers **6590**, 154–173 (2011)
7. Pham, D., Asano, S., Bolliger, M., Day, M., Hofstee, H., Johns, C., Kahle, J., Kameyama, A., Keaty, J., Masubuchi, Y., Riley, M., Shippy, D., Stasiak, D., Suzuoki, M., Wang, M., Warnock, J., Weitzel, S., Wendel, D., Yamazaki, T., Yazawa, K.: The Design and Implementation of a First-Generation CELL Processor. In: Proceedings IEEE International Solid-State Circuits Conference (2005)
8. x264. A Free H.264/AVC Encoder. http://www.videolan.org/developers/x264.html

Chapter 5
Extracting More Parallelism: the 3D-Wave

Abstract If higher performance is required, a parallel application developer might have to extract more parallelism than initially employed in the application. To illustrate this step, this chapter presents a parallel implementation of H.264 decoding on a shared-memory system that scales to a large number of cores. The application implements the dynamic 3D-Wave algorithm, which exploits intra-frame MB-level parallelism as well as inter-frame MB-level parallelism. The 3D-Wave algorithm is based on the observation that inter-frame dependencies have a limited spatial range, i.e., that motion vectors are typically short. Experimental results obtained using a simulator of a many-core architecture containing NXP TriMedia TM3270 embedded processors show that the implementation scales very well, achieving a speedup of more than 50 on a 64-core processor for a 25-frame FHD sequence.

5.1 Introduction

The performance of the Ring-Linc implementation of the 2D-Wave algorithm presented in the previous chapter is mainly limted by the memory bandwidth. The analytical results in Chapter 3 show, however, that even under ideal circumstances (no communication and synchronization overhead), the speedup of any implementation of the 2D-Wave algorithm is limited to about 22 for FHD video resolution. Therefore, in order to scale beyond a speedup of 22, not only the memory bandwidth needs to be increased, also a different algorithm needs to be employed.

In this chapter we present an implementation of the dynamic 3D-Wave algorithm on a shared-memory system, which, as was shown analytically in Chapter 3, exhibits much more parallelism than the 2D-Wave. The dynamic 3D-Wave algorithm exploits the fact that due to "inertia", motion vectors are typically short. Because of this, one does not have to wait until the entire reference frame has been decoded before starting the decoding of a frame. Instead, the decoding of a MB can start as soon as its reference area in the reference frame has been decoded and its intra-frame dependencies have been solved.

Implementing the dynamic 3D-Wave, however, is fundamentally more difficult than implementing the 2D-Wave, because the dependencies are not known statically. The dependency between a MB and its reference area is only known dynamically at runtime. To solve this problem, in this chapter we develop a *subscription mechanism* where MB decoding tasks subscribe themselves to a *kick-off list* associated with the MBs on which they depend. Thereafter, when a MB decoding task has finished, it resumes all MB decoding tasks in its associated kick-off list. The algorithm has been implemented on a simulator of a shared-memory architecture containing NXP TriMedia TM3270 embedded cores. The experimental results show that the implementation scales very well, achieving a speedup of more than 54 on a 64-core processor.

This chapter is organized as follows. To make this chapter self-contained, Section 5.2 briefly reviews the dynamic 3D-Wave algorithm. The performance of our implementation of the 3D-Wave will be compared to the performance of a 2D-Wave implementation, and in Section 5.3 we describe how the 2D-Wave can be implemented on a shared-memory system. This implementation is similar to the Task Pool (TP) implementation presented in the previous chapter, but on a shared-memory system there is no dedicated master core that manages the task pool. Instead, the task pool is a software structure in shared memory that can be accessed by all the cores. Such a shared data structure introduces a synchronization problem, because it must be accessed *atomically*. Furthermore, the 2D-Wave implementation in this chapter features an optimization called *tail submit* to improve data locality. Hence, Section 5.3 illustrates how shared-memory systems need to be programmed. The implementation of the dynamic 3D-Wave is presented in Section 5.4. Finally, experimental results are provided in Section 5.5 and conclusions are drawn in Section 5.6.

5.2 Dynamic 3D-Wave Algorithm

As previously explained in Chapter 3, the dependencies between frames in H.264 decoding are due to motion compensation and motion vector prediction. In order to decode a MB, its reference area in the reference frame has to be decoded first. Therefore, these dependencies are naturally fullfilled if frames are decoded one-by-one, as in the 2D-Wave algorithm. One does not have to wait until the reference frame has been completely decoded, however. The decoding of a MB can start as soon as its reference area has been decoded. Furthermore, because usually there is little motion between consecutive frames in a scene motion vectors are typically short (a few pixels). The decoding of a reference frame and its referencing frame, therefore, can be partially performed in parallel. This is the basic idea underpinning the 3D-Wave algorithm.

3D-Wave parallelism can be exploited either statically or dynamically. In the static approach, the worst case motion vector length is assumed and a MB decoding task is scheduled no sooner than when it is certain that its reference area has been decoded. A variation of the static approach will be considered in Chapter 7. In the dynamic approach the actual motion vectors are used and a MB decoding task is (re-)started as soon as its reference area has been decoded. In this chapter we will implement the dynamic approach.

5.3 2D-Wave Implementation on a Shared-Memory System

In this chapter, the performance of our parallel implementations of H.264 decoding will be evaluated using a simulator of a *cache-coherent* shared-memory system. Cache coherence means that the contents of a cache block in the private cache of one core is updated or invalidated when another core modifies that cache block. Furthermore, the performance of the 3D-Wave implementation will be compared to the performance of a 2D-Wave implementation. For this reason, in this section we present an implementation of the 2D-Wave algorithm on a shared-memory system.

The 2D-Wave implementation is similar to Task Pool (TP) implementation presented in the previous chapter (see Section 4.3) with the important difference that there is no longer a master core that manages the task pool. Instead, the task pool is a software data structure accessible by all cores and stored in shared memory. Cores can submit tasks to this task pool using tp_submit(fun_ptr, <params>) which, at some point, are executed by calling the function pointed to by fun_ptr. This shared data structure, however, as well as the matrix of dependency counters that is also present in the TP implementation described in the previous chapter, need to be accessed *atomically*. Atomic means that modifications to these data structures need to be indivisible, i.e., while the data structure is being modified, no other modifications may be performed. Computer architectures provide special instructions (so called read-modify-write instructions) to implement atomicity [4].

Figure 5.1 depicts pseudo-code for the MB decoding task on a shared-memory system, based on the pseudo-code given in [3]. It starts with the actual work of decoding MB(x,y). Thereafter, the dependency counters associated with the right and down-left neighbors of MB(x,y) are decremented (provided they exist), as in the TP implementation presented in the previous chapter, and if the dependency counters become zero, the corresponding MB decoding tasks are submitted to the task pool.

Unlike the previous implementation, however, the dependency counters cannot be decremented using conventional decrement operations. Otherwise, it may happen that when one MB decoding task has loaded a dependency counter in a register, another task loads the same (old, *stale*) value and one of the decrements will be lost. Listing 5.2 depicts "pseudo-assembly code" that illustrates how such a *data race* could occur. Initially, the value of dep_count[i][j] is 2. First, thread T_1 loads dep_count[i][j] in a register, say Ra. Before it is able to decrement Ra and write back the result to memory, the malicious other thread loads dep_count[i][j] in a register Rb, decrements Rb, and stores the result back to dep_count[i][j]. Therefore, the decrement of T_2 will be lost. To prevent such *race conditions*, the dependency counters are decremented using atomic decrement

```
void decode_mb(int x, int y)
{
    // ... the actual work

    // check and submit right MB
    if ( y < WIDTH-1){
        atomic_dec(dep_count[x][y+1]);
        if (dep_count[x][y+1]==0)
            tp_submit(decode_mb, x, y+1);
    }

    // check and submit down-left MB
    if ( x < HEIGHT-1 && y != 0){
        atomic_dec(dep_count[x+1][y-1]);
        if ( dep_count[x+1][y-1] == 0)
            tp_submit(decode_mb, x+1, y-1);
    }
}
```

Listing 5.1: Initial pseudo-code for MB decoding.

operations. Programmers of shared-memory systems need to be aware of such race conditions.

```
Thread T_1                              Thread T_2
Ra = load(dep_count[i][j])
                                        Rb = load(dep_count[i][j])
                                        Rb = Rb - 1
                                        dep_count[i][j] = store(Rb)
Ra = Ra - 1
dep_count[i][j] = store(Ra)
```

Listing 5.2: data race for dependency update.

As stated previously, to decode a MB, mainly data from its left and top neighbors is needed. Furthermore, if the cache line is larger than the width of a MB (16 pixels), a core also fetches data from neighboring MBs into its cache when it loads a MB. Therefore, *data locality* can be improved if the core that processes MB(x,y) also processes MB$(x,y+1)$. To exploit this locality as well as to reduce task submission overhead, the 2D-Wave implementation features an optimization called *tail submit* [3]. Pseudo-code for the MB decoding task after the tail submit optimization has been performed is depicted in Figure 5.3.

As before, after the MB has been processed, the dependency counters of its right and down-left neighbors are atomically decremented. If both of them are ready to execute, the task continues with the right MB (to increase data locality) and submits the other one to the task pool. If only one of them is ready, the task processes it and

```
void decode_mb(int x, int y)
{
    down_left_avail = 0;
    right_avail     = 0;
    do {
        // ... the actual work

        if ( x < HEIGHT-1 && y != 0){
            atomic_dec(dep_count[x+1][y-1]);
            if ( dep_count[x+1][y-1] == 0)
                down_left_avail=1;
        }
        if ( y < WIDTH-1){
            atomic_dec(dep_count[x][y+1]);
            if (dep_count[x][y+1]==0)
                right_avail=1;
        }
        // give priority to right MB
        if ( down_left_avail && right_avail)
            y++;
            tp_submit(decode_mb, x+1, y-1);
        }
        else if (down_left_avail){
            x++; y--;
        }
        else if (right_avail){
            y++;
        }
    }while (down_left_avail || right_avail);
}
```

Listing 5.3: Pseudo-code for MB decoding task after tail submit optimization.

does not submit a new task to the task pool. If there is no neighboring MB ready to be processed, the task finishes and the core requests a new task from the task pool. The astute reader will notice that this implementation of the 2D-Wave algorithm is somewhat similar to the Ring-Line implementation described in the previous chapter. The difference is that in the RL implementation the mapping of MBs to cores is completely static, whereas in the shared-memory implementation of the 2D-Wave is somewhat dynamic. Because of this, pre-fetching and pre-sending of data to the core that needs it cannot be performed in the shared-memory implementation (unless the parallel application developer is willing to do it speculatively, without being sure that the data will be needed, which is something we do not advice).

5.4 Dynamic 3D-Wave Implementation

In this section we present the implementation of the dynamic 3D-Wave algorithm. The starting point of our implementation was the 2D-Wave implementation described in the previous section. Amongst others, two issues had to be dealt with to implement the 3D-Wave algorithm. First, the reference frame buffer had to be modified. Second, and most importantly, a *subscription mechanism* where MB decoding tasks can subscribe themselves to the MB decoding tasks on which they depend had to be developed.

First, since the 3D-Wave implementation decodes multiple frames concurrently, modifications to the reference frame buffer were required. The reference frame buffer stores decoded frames that are going to be used as reference frames. In the sequential case the reference buffer can serve only one frame-in-flight. For the 3D-wave, the reference frame buffer was modified such that a single instance of the buffer can serve all frames that are concurrently being decoded.

Second, a method is needed to suspend a MB decoding task when it is discovered that its reference area has not been decoded yet and to resume it later after the reference area has been decoded. To be able to do so, a subscription mechanism has been developed where MB decoding tasks can subscribe themselves to the *kick-off lists* associated with MBs on which they depend. Specifically, let the reference MB be the MB in the bottom right corner of the reference area (including the extra samples required for fractional motion compensation, but for comprehensibility, we will from now on ignore such technical issues). Note that the reference MB is the last MB to be decoded in the reference area, due to the intra-frame dependencies. If a MB decoding task detects that its reference MB has not been decoded yet, it inserts "itself" (its coordinates as well as its frame number) in the kick-off list associated with its reference MB. After any MB has been decoded, its associated kick-off list is scanned and all MB decoding tasks in there are (re-)submitted to the task pool.

Figure 5.4 presents high-level pseudo-code for the `decode_mb` task. Its parameters are the MB coordinates x, y, frame number f, as well as a boolean `mv_pred_done` that indicates that the vector prediction kernel can be skipped since it has been performed before, and a parameter `RMB_start` that allows to skip checking certain reference MBs because they have been checked before. The first time the `decode_mb` task is called, it is called with `mv_pred_done` set to `false` and `RMB_start` set to 0.

First the motion vector prediction kernel is performed. The reference MBs are calculated (provided `mv_pred_done` is not set) and they are inserted in a list of reference MBs. Thereafter, for each reference MB in this list it is checked if it has been decoded using a 3-dimensional matrix of *ready bits*. If not, the coordinates of the current MB, the frame number, as well as the current value of the loop counter plus 1 (the next `RMB_start`) are inserted into the kick-off list `KoL` associated with the reference MB. Thereafter, the function returns, which effectively suspends the task. This insertion also needs to be performed atomically, since another task can insert itself into the same list at the same time. Only if all reference MBs have been

```
void decode_mb(int x, int y, int f, bool mv_pred_done, int
    RMB_start)
{
    if (!mv_pred_done){
        Motion_Vector_Prediction(x,y);
        RMB_List = RMB_Calculation(x,y);
    }
    for (i=RMB_start; i<RMB_List.last; i++)
    {
        (g,u,v) = get_frame_number_and_MB_coord(RMB_List.rmb[i]);
        if (!Ready[g][u][v])
        {
            Subscribe(KoL[g][u][v], x, y, f, i+1);
            return;
        }
    }
    Picture_Prediction(x,y);
    Deblocking_Info(x,y);
    Deblocking_Filter(x,y);
    Ready[f][x][y] = true;
    for (each quadruple (u, v, g, start) in KoL[f][x][y])
        tp_submit(decode_mb, u, v, g, true, start);
    // decrement dependency counters and tail submit (omitted)
}
```

Listing 5.4: Pseudo-code for the 3D-Wave.

processed, the remaining kernels (picture prediction, deblocking info, and deblocking filter) are performed.

After that, the ready bit of the current MB is set in the matrix of ready bits to inform other MB decoding tasks that this MB is ready to be used as a reference MB. In the next step, all MB decoding tasks in the kick-off list associated with the current MB are resumed, by re-submitting them to the task pool with the parameter mv_pred_done set to true and RMB_start set to the index of the next reference MB in the list of reference MBs. Finally, the dependency counters of the right and down-left neighbors of the current MB are decremented and these MBs are processed and/or submitted using the tail submit approach, similar to the 2D-Wave implementation.

Note that what we have actually implemented is a mechanism to suspend and resume a task (to be more precise: a mechanism to quit and and to start a new task from where it left off). This is similar to the functionality provided by *semaphores*, which are well-known synchronization primitives. A semaphore is an integer variable on which to operation can be performed: *down* and *up* (sometimes also called wait and signal or *P* and *V*). A down operation inspects the value of the semaphore. If it is 0, the task (or thread, or process) performing the down is suspended and the task is inserted into a list of tasks that are said to *sleep* on the semaphore. If the semaphore value is larger than 0, the task simply decrements the value and continues. An up operation also inspects the semaphore. If there is a task sleeping on this semaphore, this task is woken up and the semaphore value remains 0.

For several reasons semaphores could not be used to implement the dynamic 3D-Wave algorithm. First, they were not available in the task pool programming model we used to implement the 2D- and the 3D-Wave. The task pool model basically assumes that tasks are *non-preemptive*, i.e., when they start, they execute until their completion. Second, we have performed some preliminary experiments using semaphores provided by a threading library and discovered that a large number of semaphores would be required and that the overhead was significant. The overhead is significant because the *state* of a task (program counter, register file contents, etc.) needs to be saved when a task is suspended. In our implementation we exploit the fact that on the considered platform, it is relatively cheap to submit a new task to the task pool and a return from a function is obviously cheap also. Third, and most importantly, semaphores do not exactly provide the functionality that is required to implement the 3D-Wave. An up operation wakes up at most one task sleeping on the semaphore, but the 3D-Wave implementation requires that *all* tasks waiting on a reference MB are woken up.

5.5 Experimental Evaluation

In this section the 2D-Wave and 3D-Wave implementations described in the previous sections are experimentally evaluated. First we describe the simulator employed and the system it models. Thereafter, the experimental results are presented.

5.5.1 Experimental Setup

The 2D-Wave and 3D-Wave implementations are evaluated using a simulator of an NXP embedded many-core architecture containing TM3270 media cores. An NXP proprietary simulator based on SystemC is used to run the applications and to collect performance data. Computations on the cores are modeled cycle-accurate.

The simulator models a shared-memory cache-coherent architecture. A Network On Chip (NoC) connects the cores to each other as well as with the off-chip main memory. The memory is modeled using an Average Memory Access Time (AMAT) of 40 cycles and by taking channel and bank contention into account. When channel or bank contention is detected, the AMAT is increased. Contention in the NoC is modeled as well. The shared memory consists of 8 *banks* that can be accessed simultaneously. The simulator is capable of simulating systems with up to 64 TM3270 cores. The operating system is not simulated.

The TM3270 [5] is a VLIW-based media-processor based on the Trimedia architecture. In a VLIW processor each instruction consists of several *slots* and the operations in these slots are executed concurrently. The TM3270 is targeted at the requirements of multi-standard video processing at standard resolution and the associated audio processing requirements for the consumer market. The architecture supports VLIW instructions with five *guarded* issue slots. Guarded means that each slot contains a predicate or guard and the operation in the slot writes back its result to the register file only when the predicate evaluates to true. This is a common technique to increase Instruction-Level Parallelism (ILP) in VLIW processors. The pipeline depth varies from 7 to 12 stages. Address and data words are 32 bits wide and the register file consists of 128 32-bit registers. 2×16-bit and 4×8-bit SIMD instructions are also supported. SIMD instructions are instructions that process several data elements (in this case 2 16-bit elements are 4 8-bit elements) simultaneously and are targeted at exploiting Data-Level Parallelism (DLP). The TM3270 processor can run at up to 350 MHz, but in this work the clock frequency is set to 300 MHz. To generate code for the TM3270 the NXP TriMedia C/C++ compiler version 5.1 is used.

Each core has a private level-1 (L1) data cache and can copy data from other L1 data caches through 4 channels. The L1 data cache parameter values are as follows: 64KB size, 64B line size, 4-way set-associative, write back, and write-allocate. Instruction caches are not modeled, since typically they have high hit rates. The contents of the private L1 data caches are kept coherent using the MESI cache coherence protocol. Very briefly, in this protocol each cache line can be in one of four states: *Modified*, *Exclusive*, *Shared*, or *Invalid*. Modified means that this is the only copy of the block in the system and this copy has been modified. Shared means that there can be other copies of the block. This block can therefore be read without informing other cores, while in case of a write, all other copies need to be invalidated. Invalid means that the block is not present because it has been invalidated by a write from another core. Exclusive means that this is the only copy of the block and that this block has not been modified since it was loaded from memory. Unlike shared blocks, it can therefore be written without having to invalidate other copies.

As explained before, the programming model is that of a task pool. A library provides primitives for submitting and requesting tasks to/from the task pool as well as synchronization primitives. Each core runs a thread that implements the so-called Fetch&Execute cycle: request a task from the task pool, execute the task, and repeat. The overhead of requesting and submitting tasks is low. For example, the time to request a task is less than 2% of the average MB decoding time.

The experiments focus on the baseline profile of the H.264 standard. This profile only supports I- and P-frames and every frame can be used as a reference frame. Because every frame can be a reference frame, this profile corresponds to the worst case for the 3D-Wave strategy, since frames are never completely independent. The video sequences were encoded using the X264 encoder [6] using the following options: no B-frames, at most 16 reference frames, weighted prediction, hexagonal motion estimation algorithm with a maximum search range of 24 pixels, and one slice per frame. The experiments were performed using all four video sequences from the HD-VideoBench [1] (Blue_Sky, Rush_Hour, Pedestrian, and Riverbed) at FHD resolution. To save space, however, only the results for the Rush_Hour sequence at FHD will be presented. These results correspond to the median of all results and in all cases, the results for the other sequences differ less than 5%.

As mentioned before, the results presented in this and the previous chapter do not include the time to entropy decode the input bitstream, since there is hardly any parallelism in the entropy decoding of one frame (but different slices/frames can be decoded in parallel, as we will show in later chapters). Instead we assume that a hardware accelerator is available to perform entropy decoding. In the simulations, entropy decoding was performed off-line. The performance requirements of an entropy decoder accelerator will be analyzed in Chapter 6.

5.5.2 Experimental Results

To evaluate the 2D-Wave and 3D-wave implementations, the time it takes to decode one second (25 frames) of each video sequence was measured. Longer sequences could not be simulated due to constraints of the simulator. From 1 up to 64 cores were simulated and the number of cores had to be a power of 2. More cores could not be used due to limitations of the simulator.

5.5.2.1 Performance Results

Figure 5.1 depicts the speedup achieved by the 2D-Wave and 3D-Wave implementations as a function of the number of cores. Here, speedup is defined as the speedup of a parallel implementation running on p cores over the 2D-wave code on a single core. In order to be able to directly compare the performance of the 3D-Wave implementation to the performance of the 2D-Wave implementation, the speedup of the 3D-Wave implementation over the 2D-Wave implementation executed on a single core (the curve labeled "3D vs 2D") is also shown.

The figure clearly shows that the 3D-Wave implementation scales almost perfectly up to 64 cores, achieving a speedup of 50.4. The 2D-Wave implementation, on the other hand, scales reasonably well up to 16 cores (the number of cores considered in the previous chapter), but thereafter, its efficiency decreases rapidly. It attains a speedup of 15.4 on 64 cores. Clearly, the 2D-Wave algorithm does not exhibit sufficient parallelism to scale to 32 cores or beyond.

Focussing in on the speedup of the 3D-Wave over the 2D-Wave running on a single core, we observe that this speedup is slightly lower that the speedup of the 3D-Wave over itself. This is because on a single core, the 2D-Wave implementation is 8% more efficient than the 3D-Wave implementation. The reason for this is that the 3D-Wave incurs more runtime overhead. The subscription/kick-off mechanism it requires, although efficient, is not for free. However, already at 2 cores, the performance of the 3D-Wave surpasses the performance of the 2D-Wave implementation.

As stated before, the 3D-Wave implementation achieves a speedup of 50.4× on 64 cores, corresponding to an efficiency of almost 85% (efficiency is defined as the

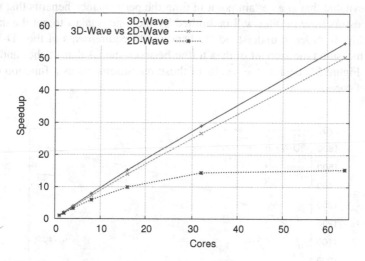

Fig. 5.1: 2D-Wave and 3D-Wave speedups for the 25-frame Rush_Hour sequence in FHD resolution.

ratio of the speedup over the number of cores). The main reason why the speedup is less than perfect is that due to memory size restrictions, very short video sequences (25 frames, 1 second) had to be used as input. Using longer input sequences increases the speedup of the 3D-Wave because the ramp-up and ramp-down phases become negligible. To proof this, the article on which this chapter is based [2] also presented results for a 100-frame sequence in SD resolution. Because SD frames are much smaller than FHD frames, a longer SD sequence could be simulated. On 64 cores, using a 100-frame sequence instead of a 25-frame sequence increases the speedup of the 3D-Wave by 13% from 49.3× to 55.7×. The 3D-Wave implementation only incurscw the ramp-up overhead for the first frame and the ramp-down for the last frame. The 2D-Wave, on the other hand, incurs this ramping overhead for each and every frame.

The main reason why the 3D-Wave achieves higher scalability than the 2D-Wave is "simply" because it exhibits much more parallelism. This parallelism is not nearly exhausted at 64 cores. In the 2D-Wave, on the other hand, the parallelism is exhausted at 20-25 cores for FHD resolution. While its parallelism increases with the resolution, it requires an extremely high resolution for the 2D-Wave to scale to 64 cores or beyond.

5.5.2.2 Bandwidth Requirements

The memory bandwidth requirements is major concern for architects of multi-core processors. Since the number of pins that can be used to connect off-chip memory to the processor increases at a much smaller rate than the number of transistors, it seems inevitable that at a certain point in time the performance benefits that multi-core processors have to offer will be limited by the time it takes to get the data on and off chip. In order to understand the bandwidth requirements of the 3D-Wave, we have measured the amount of data traffic between the L1 data caches and main memory. Figure 5.2 depicts the results of these measurements as a function of the number of cores.

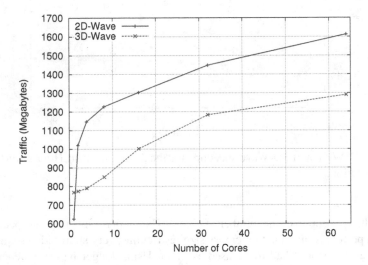

Fig. 5.2: Amount of data traffic between the L1 caches and main memory for the 2D-Wave and the 3D-Wave for the 25-frame Rush_Hour sequence.

It can be seen that increasing the number of cores increases the amount of data that needs to be transferred between the L1 data caches and main memory. This implies that data is not brought once onto the processor, stays there to be processed, and then the results are written back, but that cache conflict misses occur, causing data to be written back and re-loaded from memory. This effect is most visible for the 2D-Wave when moving from 1 to 2 cores. Apparently, data re-use that could be exploited on a single core is lost when moving to 2 cores. Thereafter, the amount of data traffic increases more slowly, and is likely due to *sharing misses* that occur in cache-coherent systems when the data read by one core is invalidated by another core, forcing the first core to refetch it when it needs it again. The number of sharing misses typically increases with the number of cores. Interestingly, and somewhat unexpectedly, is that for the 3D-Wave the amount of data traffic increases slower than for the 2D-Wave. The reason for this is that the largest re-use of data in the MB decoding phase of H.264 decoding is between a MB and its reference area. In the 2D-Wave the tasks producing these two pieces of data are separated by a barrier, since a frame needs to be decoded completely before the decoding of the next frame can start. In the 3D-Wave these tasks can be scheduled closer together in time, and indeed they are, since when a reference MB has been decoded it kicks off all MBs that depend on it. Tasks in the task pool are scheduled in a first-in-first-out order, but due to the above, it may be that a last-in-first-out order performs even better, requires even less data traffic. This, however, has not been investigated.

5.6 Conclusions

In this chapter we have described implementations of the 2D-Wave and the 3D-Wave on cache-coherent shared-memory systems. The 2D-Wave implementation is similar to the Task Pool (TP) implementation presented in the previous chapter, but unlike the TP implementation, the implementation presented in this chapter does not feature a master core that automatically provides synchronized access to the shared data structure. Instead, the task pool is stored in shared memory and must be accessed using atomic operations. The 3D-Wave implementation requires an intricate subscription/kick-off mechanism to solve the dependencies between a MB and its reference MBs. While the required functionality is similar to the well-known synchronization primitive semaphore, it is not an exact match and, therefore, we had to develop our own mechanism based on detailed knowledge of the "nature" of the dependencies, emphasizing again the importance of the step "understanding the application". Experimental results obtained using a simulator of an embedded shared-memory multicore architecture show that the 3D-Wave is indeed capable of scaling to a much larger number of cores than the 2D-Wave, as was predicted by the analytical results presented in Chapter 3. This might seem evident, but architectural constraints (e.g., memory bandwidth limitations, tasking overhead, synchronization cost, etc.) often cause parallel applications to scale worse in practice than predicted in theory, as was also shown for the TP implementation in the previous chapter. Because the evaluation platform considered in this chapter has a very efficient implementation of the task pool, the 3D-Wave implementation achieves a speedup of more than 50 on 64 cores, corresponding to an efficiency of almost 85%. Furthermore, the main reason why the speedup is less than perfect is that, due to the short sequences (25 frames) used for the simulations in which the ramping effects at the begining and at the end of the sequence are more noticeable. For real sequences, with thousands of frames, the ramping effects will be negligible.

Several conclusions about how to program many-core architectures can be drawn on the basis of the work presented in this chapter. First, in order to scale to a large number of cores, all parallelism exhibited by the application should be exploited. This was clearly shown in this chapter, because the 3D-Wave implementation achieves a much higher speedup on 64 cores than the 2D-Wave implementation. The scalability of the 2D-Wave, however, scales with the resolution, and for higher resolution higher performance is required. Second, while cache-coherent shared-memory systems provide a convenient programming model because data is communicated implicitly between task via the shared memory, the parallel programmer still has to be aware of the location of data in order to improve data locality. The tail submit optimization described in this chapter provides evidence for this. If possible and provided it does not reduce parallelism, a task should be executed on the same core as the task that produces the data that it requires. Third, while contemporary parallel programming languages/libraries support several synchronization constructs (locks, semaphores, atomic operations, etc.), they do not always provide the functionality for the synchronization problems encountered. This is illustrated in this chapter by the subscription/kick-off mechanism we had to develop in order to solve the dependencies between a MB and its reference MBs.

As mentioned before, the results in this and the previous chapter do not include the time required to perform entropy (CABAC) decoding. Basically we assumed that a hardware accelerator that performs entropy decoding is available, so that this stage of H.264 decoding does not constitute a bottleneck. The validity of this assumption will be analyzed in the next chapter.

References

1. Alvarez, M., Salami, E., Ramirez, A., Valero, M.: HD-VideoBench: A Benchmark for Evaluating High Definition Digital Video Applications. In: IEEE International Symposium on Workload Characterization (2007). http://personals.ac.upc.edu/alvarez/hdvideobench/index.html
2. Azevedo, A., Juurlink, B., Meenderinck, C., Terechko, A., Hoogerbrugge, J., Alvarez, M., Ramirez, A., Valero, M.: A highly scalable parallel implementation of h.264. Transactions on High-Performance Embedded Architectures and Compilers **4**(2) (2009)
3. Hoogerbrugge, J., Terechko, A.: A Multithreaded Multicore System for Embedded Media Processing. Transactions on High-Performance Embedded Architectures and Compilers **6590**, 154–173 (2011)
4. Taubenfeld, G.: Synchronization Algorithms and Concurrent Programming. Prentice Hall (2006)
5. van de Waerdt, J., Vassiliadis, S., Das, S., Mirolo, S., Yen, C., Zhong, B., Basto, C., van Itegem, J., Amirtharaj, D., Kalra, K., et al.: The TM3270 Media-Processor. In: Proceedings of the 38th International Symposium on Microarchitecture, pp. 331–342 (2005)
6. x264. A Free H.264/AVC Encoder. http://developers.videolan.org/x264.html

Chapter 6
Addressing the Bottleneck: Parallel Entropy Decoding

Abstract In the previous chapters we mainly focused on generally the most time-consuming phase of H.264/AVC decoding, the macroblock reconstruction phase. There is another phase, however, the entropy decoding phase, that takes a significant amount of time. Therefore, Amdahl's law teaches us that this phase also needs to be parallelized and in this chapter we show how this can be done. In order to be able to do so, however, dependencies that result from reusing sequential legacy code need to be eliminated. Experimental results gathered on a shared-memory multicore/multiprocessor system show that the presented parallel entropy-decoder scales well, but also that the obtained speedup depends in the time taken by the entropy decoding phase, due to the memory bandwidth bottleneck.

6.1 Introduction

In the previous chapters, in which we focused on the macroblock (MB) reconstruction phase of H.264 decoding, we have been cheating a little. While the MB reconstruction phase is in general certainly the most time-consuming phase, there is another phase, the entropy decoding phase, that requires a significant amount of time. Cheating is not the right word, because we did write that we assumed that entropy decoding would be performed by an accelerator and, indeed, commercial H.264 decoders often contain a hardware IP block that performs entropy decoding. But commercial H.264 decoders focus on at most FHD (1080p) resolution with a throughput of 25 or 50 frames per second (fps) and a relatively low bitrate. For higher bitrates, resolutions and/or higher throughput, the performance of the accelerator would have to scale accordingly. But can this really be achieved?

We are thus confronted with Amdahl's malicious law. Informally, *Amdahl's law* applied to H.264 decoding tells us that if the entropy decoding takes only 5% of the total execution time and we do not parallelize it or accelerate it otherwise, the speedup will be limited to 20, no matter how many cores we employ in the MB reconstruction phase. A bit more formally, Amdahl's law states that if a fraction s of an application's execution time is sequential and the remaining part $1 - s$ is perfectly parallelizable, then the maximum speedup S_{max} on p cores is at most

$$S_{max} = \frac{1}{s + \frac{1-s}{p}} \qquad (6.1)$$

Although Amdahl's law seems simple, it is surpirsing that some students get it wrong during a test.

After focusing on the most time consuming phase of H.264 decoding, we take the next step: addressing the (entropy decoding) bottleneck. Jestingly, this step might also be called "defeating Amdahl's law". We remark, however, that this step might be necessary several times during the parallel-application-design-cycle. In fact, when we moved from the centralized task pool implementation to the more distributed Ring-Line implementation in Chapter 4, this step was also performed.

This chapter is organized as follows. In Section 6.2 we profile a sequential H.264 decoder in order to determine the percentage of time each phase or kernel takes. Based on these results and using a small generalization of Amdahl's law, we determine the speedup that an entropy decoding accelerator should provide in order to obtain a scalable parallel H.264 decoder. A work of science is not a detective[1] and we already disclose the conclusion: it is very unlikely that a hardware accelerator can ever provide this level of performance. Therefore, to obtain a scalable implementation, the entropy decoding phase also needs to be parallelized, and we show how it can be done in Section 6.3. In addition, we discuss how entropy decoding can be decoupled from the MB reconstruction phase, how dependencies that result from recycling legacy sequential codes can be eliminated, and how the entropy decoding and MB reconstruction stages can be interfaced. Experimental results of our parallel entropy decoder on a shared-memory system are presented in Section 6.4. Finally, conclusions are drawn in Section 6.5. Here we remark that in this chapter, as we do in all other chapters, we focus on Context Adaptive Binary Arithmetic Coding (CABAC), because it achieves higher compression than the other entropy coding technique included in the H.264 standard, Context Adaptive Variable Length Coding (CAVLC), and because it is widely used in HD applications.

6.2 Profiling and Amdahl

In order to analyze the speedup required from a hardware entropy decoder, a state-of-the-art sequential H.264 decoder (FFmpeg [3]) is profiled for different videos, resolutions, and coding options. For this, the application is divided into four stages: the front-end (reading and parsing the bitstream), entropy (CABAC) decoding, MB reconstruction, and Motion Vector Prediction. The motion vector prediction phase includes also reference index prediction, and intra-mode prediction. Motion vector prediction is separated from the other stages because it is the source of additional

---

[1] Quote due to some scientist whose name we forgot.

data dependencies, and later (Section 6.3.2) we will show how to remove them (or, more exactly, how to move them).

Two video resolutions and frame rates are used: FHD at 25 fps (1080p25) and QHD at 50 fps (2160p50). For 1080p25 a complete video sequence is constructed by joining three different videos: *BlueSky*, *PedestrianArea*, and *Riverbed* [1]. For 2160p50 two videos are joined: *CrowdRun* and *ParkJoy* [4]. By combining different videos into one, a real video with different scenes is emulated. Table 6.1 shows the main properties of the selected sequences.

| Video | Resolution | Frame | Bitrate [Mbps] | | | | |
name	[pixels]	rate	crf22	crf27	crf32	crf37	intra
BlueSky+Pedestrian+Riverbed	1920 × 1080	25	12.2	6.17	3.28	1.72	51.2
CrowdRun+ParkJoy	3840 × 2160	50	109	48.3	23.6	12.0	405

Table 6.1: Properties of the video sequences.

Five encoding modes are analyzed: four with constant quality encoding and with intra-only. To produce constant quality encoding videos, the X264 open-source encoder [9] was employed with four different values of the Constant Rate Factor (CRF): 22, 27, 32, 37. A smaller CRF results in higher bitrate and higher quality. CRF-22 is representative of high quality, high definition applications, and the other extreme, CRF-37, is typical for medium- to low-quality Internet video streaming applications. Intra-only encoding is representative of professional video systems such as video editing and high quality camera equipment. This mode uses a constant bitrate with 50Mbps for 1080p25 and 400Mbps for 2160p50. If you do not understand the above in detail, please do not worry. The main message of this paragraph is that we did our job right by considering different and representative options, which is something researchers have to do to convince their peers that their articles should be accepted and that their research should be funded.

Figures 6.1a (1080p25) and 6.1b (2160p50) breaks down the average decoding time per frame into the time taken by the front-end, entropy decoding, MB reconstruction, and the motion vector prediction stages. Three main conclusions can be drawn from these results. First, CABAC decoding, MB reconstruction, and motion vector prediction together consume on average 99.5% of the total execution time. Second, entropy decoding expends a significant fraction of the total execution time. On average, over all videos, resolutions, and coding options, it consumes 37.6% of the time it takes to decode a frame. Here it needs to be mentioned that the MB reconstruction and mode prediction stages have been vectorized using SIMD instructions, while these instructions cannot be used to accelerate the CABAC decoding stage, since it does not exhibit data-level parallelism at which SIMD instructions are targeted. The third main conclusion that can be drawn from the results is that the fraction of time taken by each stage is not constant but varies substantially depending on the coding option. The fraction of the total execution time taken by CABAC decoding ranges, on average over the two resolutions, from 20% for CRF-37 to 72% for Intra-only.

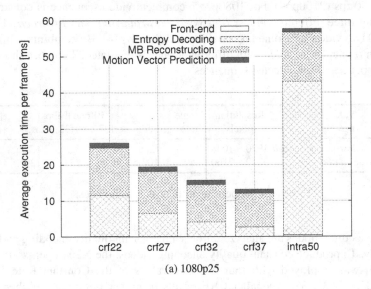

(a) 1080p25

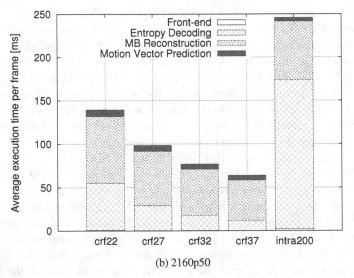

(b) 2160p50

Fig. 6.1: Profiling of a sequential H.264 decoder for different videos, resolutions, and coding options.

From the average fraction taken by entropy decoding, a small variation of Amdahl's law, and the targeted scalability, the performance required from a hardware entropy decoder can be estimated. Let's suppose a hardware entropy decoder can boost the entropy decoding stage by a factor of f. Then Amdahl's law can be rewritten as

$$S_{max} = \frac{1}{\frac{s}{f} + \frac{1-s}{p}}, \tag{6.2}$$

where s is the fraction of time taken by the entropy decoding stage and p, as before, the number of cores.

Let's fill out some numbers. As stated above, on average entropy decoding consumes 37.6% of the total execution time, so $s = 0.38$. Actually, taking the average is a bad idea because a hardware IP block must be designed for the worst case, but here we ignore this for simplicity. In Chapter 5 we achieved a speedup for the MB reconstruction phase using the 3D-Wave algorithm of 50 on 64 cores. If we target, say 80%, of this performance for the entire application, Eq. (6.2) becomes:

$$0.8 \times 50 = \frac{1}{\frac{0.38}{f} + \frac{0.62}{64}} \tag{6.3}$$

Using high-school algebra, this equation can be solved for f to obtain $f = 24.8$. Thus, in order to obtain 80% of the speedup we obtained in Chapter 5, the entropy decoding stage needs to be accelerated by a factor of 24.8. No hardware block, well, let's be a bit careful and write that "it is very unlikely that a hardware block can be constructed that accelerates entropy decoding that much", especially since it exhibits hardly any parallelism, as we will show later. Indeed, it has been demonstrated that hardware acceleration is not able to provide the required entropy decoding performance for high quality and high bitrate applications [8]. This observation is based on the fact that even with application specific hardware, it is not possible to increase the CABAC decoding throughput significantly [10, 5, 7]. Therefore, to obtain a scalable parallel implementation of H.264 decoding, the entropy decoding stage also needs to be parallelized. It is inevitable.

6.3 Parallelizing CABAC Entropy Decoding

In order to be able to parallelize the CABAC entropy decoding stage, it is first necessary to understand the algorithm a little and to identify the data dependencies (remember the steps "understand the application" and "discover the parallelism"? Apparently, we have entered a cycle in our design process. Let's hope we will find our way out.). After re-taking these two steps, we show how CABAC decoding can be parallelized based on exploiting frame-level parallelism.

6.3.1 High-Level Overview of CABAC

Springer, we have a problem. We need to explain the CABAC algorithm but we don't have the space to do it in detail. Rather than confronting the reader with many terms and acronyms that probably will have no meaning to him or her, in this section we only give a very high-level of CABAC, hoping that the reader will trust us that CABAC can only be parallelized at the slice/frame-level. Readers who are interested in knowing the details can have a look at a 17-page article [6].

CABAC stands for Context Adaptive Binary Arithmetic Coding. We explain this from right to left. Arithmetic coding means that each symbol is not encoded separately, as in the perhaps more well-known Huffman coding, but entire messages (here: frames or slices) are encoded in a single number between 0.0 and 1.0 in a variable precision representation. Binary means that only binary symbols (so-called bins) are encoded. Non-binary syntax elements (e.g. transform coefficients, motion vector differences, etc) are converted to bins using a binarization step in the encoder. Finally, context adaptive expresses that the symbol probabilities, stored in context models, are adapted during encoding.

A very high-level view of a CABAC decoder is depicted in Figure 6.2. The Arithmetic Decoding block uses a shifting window of bits from the bitstream (the number between 0.0 and 1.0) to decode the binarized syntax elements, one bin at a time. Based on the outputted bins, the Context Modeler is updated. After a complete binary syntax element is detected, Inverse Binarization is performed, which produces the syntax elements that are the input for the MB reconstruction stage. The Context Modeler needs to be updated before the next bin can be arithmetically decoded, and this *feedback loop* is one of the main reasons why CABAC decoding of a single message cannot be parallelized.

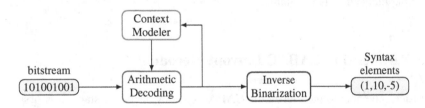

Fig. 6.2: CABAC decoder

We hope that with this very brief and very high-level description of CABAC decoding, we have convinced you a little that it is not possible to parallelize it at any level below the frame-/slice-level. Fortunately (from a "scalability point of view"), it can be parallelized at the frame-level, as will be explained in the next section.

6.3.2 Frame-level Parallelization of CABAC

An H.264 bitstream contains special *markers* (bitstrings that are not used as codewords), called Start Code Prefixes, to signal the start of a slice. In this book we assume the worst case one slice per frame, since using many slices increases the bitrate and since many videos employ 1 slice/frame, and therefore refer to these markers are frame markers. These markers can be used for fast forwarding, but here we exploit them to parallelize CABAC decoding as follows. There is a *read* thread that reads the bitstream from disk and detects frame markers. As soon as it finds a frame marker, it passes (a pointer to) the frame on to one of several *entropy decoding threads*. Provided the read thread is faster than the entropy decoding threads (it is), several frames are entropy decoded in parallel, at least partially overlapped in time.

We integrated this approach in the state-of-the-art sequential H.264 decoder FFMPEG [3]. While doing so, however, we discovered that in FFMEG, as well as in several other sequential implementations of H.264 decoding, the motion vector prediction stage is included in the CABAC stage. This improves the cache locality and hence performance of the sequential implementation because the MBs need to be loaded into the cache only once, but introduces false dependencies as explained next.

6.3.2.1 Eliminating False Dependencies

Almost always parallel applications are developed starting from a sequential implementation and our case is no different. Sequential, *legacy* codes, however, have not been designed with parallelism in mind and this may introduce dependencies that are not actually needed. We refer to this problem as the *legacy code problem* and to such dependencies as *false dependencies*, since they are similar to *false sharing misses* that occur in cache-coherent systems when the word written by one core is in the same cache block as the word read by another core.

The H.264/AVC standard includes an inter-prediction mode called *direct mode*. In this mode some MBs in B-frames do not have their own motion vectors but instead reuse the motion vectors of the co-located MB in the closest reference frame. This is done to squirt every possible bit from the bitstream and is useful if there is little motion in the video. If motion vector prediction is included in the CABAC stage, however, it causes a data dependency in the CABAC stage which needs to be obeyed by the parallelization scheme. This dependency is not a CABAC dependency, though, but a dependency that needs to be obeyed in the motion reconstruction stage (and can be fulfilled by the 2D-Wave as well as the 3D-Wave algorithms). Therefore, from the point of view of the CABAC stage, this dependency is a false dependency and is due to a legacy issue.

Figure 6.3 illustrates the problem. In this figure some MBs in the two B-frames use the motion vectors of the co-located MB in the previous reference frame, and the resulting data dependencies are indicated by dashed arrows. Therefore, if the

Frame type

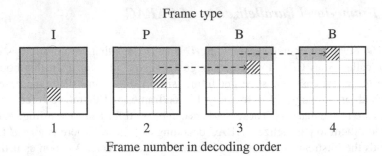

Frame number in decoding order

Fig. 6.3: False dependencies in the CABAC stage introduced by including Motion Vector Prediction in the CABAC stage. Colored MBs have been entropy decoded. Hashed MBs are currently being decoded in parallel

frames are entropy decoded in parallel, it needs to be ensured that the co-located MB is decoded before the MB that reuses its motion vectors. This is the solution we employed in the article on which this chapter is based [2]. Without going into too many details, there we proposed a parallelization strategy called *Entropy Ring*. In this strategy we ensured that the thread that entropy decodes Frame i always stays a little bit (at least one MB) behind the thread that entropy decodes Frame $i - 1$, thereby guaranteeing that the dependencies are obeyed. This implies, however, that the entropy decoding threads need to be synchronized, incurring synchronization overhead. Here we do better.

The dependencies can be removed completely from the CABAC stage by *restructuring* the application. (Perhaps yet another step that we have not included explicitly in our design process but may be necessary.) This may sound simple, but encompasses more than just moving 10 lines of code, but we will spare you the details. By decoupling CABAC from the motion vector prediction stage, the dependencies are moved to the MB reconstruction stage, where they belong. Furthermore, this can be done safely because the dependencies that need to be obeyed during the MB reconstruction stage cover the dependencies between direct-mode MBs and their co-located MBs. A potential drawback of this decoupling is, however, that it may (slightly) increase the sequential execution time. This is a trade-off that needs to be made, but since multi-core has become / is becoming the standard, we believe that in the future a higher priority should be given to parallelism rather than to sequential execution time.

6.3.2.2 Decoupling Entropy Decoding and Macroblock Reconstruction

In order to entropy decode multiple frames in parallel, the CABAC stage needs to be decoupled from the MB reconstruction stage. For communicating CABAC results to the MB reconstruction stage a frame buffer of syntax elements is required.

For each MB in a frame a minimum set of syntax elements has to be stored in the frame buffer. Then, for each CABAC frame decoded in parallel, one frame buffer of syntax elements is necessary. As a negative side effect, the memory usage of the application is increased and the data locality of the sequential version is decreased. This is the price to pay for a frame-level parallelization of entropy decoding, but in H.264 decoding there is no other alternative.

6.4 Experimental Evaluation

In this section we present some results of our decoupled parallel CABAC decoder on a shared-memory system. Before we present the results in Section 6.4.2, we first briefly describe the experimental setup (baseline implementation, input video sequences, and evaluation platform).

6.4.1 Experimental Setup

The parallel CABAC decoder has been implemented by restructuring our parallel H.264 decoder which, in turn, has been based on FFMPEG a long time ago [3]. In order to measure the scalability of the CABAC stage separately, the MB reconstruction stage has been deactivated (commented out). A read thread reads the bitstream from disk and multiple CABAC decoding threads entropy decode different frames in parallel, but the output of the CABAC decoding threads is not consumed by the MB reconstruction stage.

As input video sequences the same videos as described in Table 6.1 in Section 6.2 were employed. As remarked in that section, each video has been constructed by concatenating several different videos in order to emulate real videos consisting of multiple scenes.

The evaluation platform is a 40-core cc-NUMA system. cc-NUMA stands for *cache-coherent Non-Uniform Memory Access*. Very briefly, it is a parallel architecture where the shared memory corresponds to the aggregate of all local memories associated with each core. Therefore, the memory access time is non-uniform because a core can access its local memory faster than it can access the local memories of other cores (remote memories).

Our cc-NUMA system consists of four sockets and each socket contains 10 cores. Furthermore, each core is 2-way SMT SMT stands for *Simultaneous Multithreading* and 2-way SMT means that each core can execute more or less two threads in parallel. More or less here signifies that, although a core can execute two threads simultaneously, these threads need to share certain resources, which usually results in lower per-thread performance than when a thread is run in isolation. Each socket in our platform contains a 30MB level-3 (L3) cache that is shared between the cores on the socket, as well as 2 memory controllers. For those who would like to know

System		Software	
Processor	Intel Xeon E7-4870	Base H.264 decoder	FFmpeg
ISA	X86-64	Compiler	GCC-4.4.5
μ-architecture	Westmere	Optimization level	-O2
Sockets	4	Operating system	Debian 6.0.4
Cores/socket	10	Kernel	3.1.1
Threads/core	2		
Clock frequency	2.40 GHz		
Shared L3 cache	30MB / socket		
TurboBoost	disabled		

Table 6.2: Evaluation platform.

the nitty-gritty details, additional information about the hardware as well as the software used for the evaluation is provided in Table 6.2.

6.4.2 Experimental Results

Figure 6.4a and 6.4b depict the speedup achieved by the parallel CABAC decoder for 1080p25 and 2160p50 videos, respectively, as a function of the number of threads/cores and for different encoding modes. The speedup is relative to the performance of the parallel code running on a single core. We mention this explicitly because sometimes we as well as other researchers report speedup relative to the performance of the sequential code, which does not include certain control overhead necessary in a parallel implementation. The number of threads/cores is varied from 1 to 40. Each run was repeated three times and the reported results are the average of these runs.

As usual, several conclusions can be drawn from the figures. First, for intra-only encoded videos the speedup is almost perfect, since it increases almost linearly with the number of threads (at least 36.0 for 40 cores). If a video is intra-encoded, the CABAC stage takes a significant amount of time (more than 70% of the total execution time, cf. Figure 6.1). In this case the arithmetic decoding stage (cf. Figure 6.2 performs more operations to decode the syntax elements, which in turn, implies that the application is not memory-bound. Memory-bound means that the speedup is restricted by the memory bandwidth, i.e., it takes more time to read the data from memory than it takes to compute the result, a significant concern for multi-core applications. Second, the higher the Constant Rate Factor (CRF), the lower the speedup. For CRF-37, the speedup reaches its peak at around 25 cores for both 1080p25 and 2160p50 videos. As explained in Section 6.2, a higher CRF results in lower bitrate and lower quality. Furthermore, as shown in Figure 6.1, if the CRF is increased, the time taken by the CABAC stage reduces, while the required memory bandwidth stays the same or even increases a little. Thus we are facing the *memory bandwidth bottleneck*, yet another bottleneck that parallel application developers need to worry about.

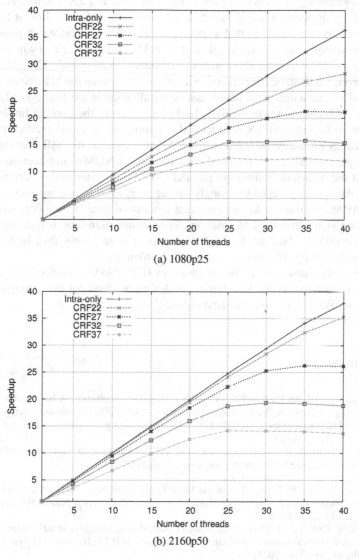

(a) 1080p25

(b) 2160p50

Fig. 6.4: Speedup of the parallel CABAC decoder, for different video resolutions, frame rates, and encoding modes.

6.5 Conclusions

In this chapter we have taken the fifth step in our parallel-application-design-process: addressing the (CABAC) bottleneck. By profiling a sequential H.264 decoder and by applying a small variation of Amdahl's malicious law, it has been shown that it is very unlikely that a hardware CABAC accelerator can provide the performance required to obtain a scalable parallel H.264 decoder as a whole. Therefore, the CABAC decoding phase needs to be parallelized as well. CABAC decoding, however, needs to be parallelized in a different way than the MB reconstruction phase, and we have described how it can be parallelized at the frame-level, i.e., by decoding different frames in parallel. To do so, however, the parallel implementation needs to be significantly restructured, by moving the mode prediction stage, a stage that is normally included in the CABAC stage, to the MB reconstruction phase. Experimental results obtained on a 40-core cc-NUMA architecture demonstrate that the resulting frame-level parallel CABAC decoder scales very well, but also that the speedup depends strongly on the *computation-to-communication ratio*. If CABAC decoding takes a significant amount of time (which depends on the type of the video), then the computation-to-communication ratio is high and nearly perfect speedup is obtained. If CABAC decoding is relatively fast, however, the speedup is bound by the *memory bandwidth bottleneck*.

Now that we know how all major stages in H.264/AVC decoding can be parallelized, it is time to take the last step in our design process: put everything together or re-iterate. We will do so in the next chapter.

References

1. Alvarez, M., Salami, E., Ramirez, A., Valero, M.: HD-VideoBench: A Benchmark for Evaluating High Definition Digital Video Applications. In: IEEE International Symposium on Workload Characterization, pp. 120–125 (2007). URL http://people.ac.upc.edu/alvarez/hdvideobench
2. Chi, C.C., Juurlink, B.: A QHD-capable Parallel H.264 Decoder. In: Proceedings of the International Conference on Supercomputing, pp. 317–326 (2011)
3. FFmpeg. http://ffmpeg.org
4. Haglund, L.: The SVT High Definition Multi Format Test Set. Tech. rep., Sveriges Television (2006). ftp://vqeg.its.bldrdoc.gov/HDTV/SVT_MultiFormat/SVT_MultiFormat_v10.pdf
5. Kim, C.H., Park, I.C.: High speed decoding of context-based adaptive binary arithmetic codes using most probable symbol prediction. In: Proceedings IEEE International Symposium on Circuits and Systems (2006)
6. Marpe, D., Schwarz, H., Wiegand, T.: Context-based Adaptive Binary Arithmetic Coding in the H.264/AVC Video Compression Standard. IEEE Transactions on Circuits and Systems for Video Technology **13**(7), 620–636 (2003)
7. Osorio, R., Bruguera, J.: An FPGA Architecture for CABAC Decoding in Manycore Systems. In: International Conference on Application-Specific Systems, Architectures and Processors, pp. 293–298 (2008)
8. Sze, V., Demircin, M.U., Budagavi, M.: Cabac throughput requirements for real-time decoding. Tech. Rep. VCEG-AJ31, ITU-T Q.6/SG16 VCEG (2008)

9. x264. A Free H.264/AVC Encoder (2011). `http://www.videolan.org/developers/x264.html`

10. Yi, Y., Park, I.C.: High-Speed H.264/AVC CABAC Decoding. IEEE Transactions on Circuits and Systems for Video Technology **17**(4), 490–494 (2007)

Chapter 7
Putting It All Together: A Fully Parallel and Efficient H.264 Decoder

Abstract It previous chapters we have presented efficient and scalable parallelization strategies for different parts (stages) of H.264/AVC decoding. To obtain a fast
and scalable parallel decoder, however, *all* stages need to be parallelized. In this
chapter we will take the final step in our parallel application design process by
putting together everything we learnt in the previous chapters in order to realize
a highly efficient and scalable parallel application. Specifically, in this chapter we
combine pipelining parallelism with data-level in the form of macroblock-level parallelism to obtain a fully parallel H.264 decoder that is optimized for core counts
of future multicore systems and emerging video decoding scenarios. The presented
implementation is evaluated on a 40-core cc-NUMA system using 1080p25 and
2160p50 video sequences.

7.1 Introduction

On contemporary high-performance processors, optimized single-threaded H.264
decoders or implementations with a few threads meet the performance requirements of 1080p25 videos. Since the release of the H.264/AVC standard in May
2003, single-threaded processor performance has improved sufficiently for real-time
decoding of 1080p25 video sequences. So the developers of the H.264/AVC standard were among the last developers who enjoyed the free software lunch. A single
core/thread cannot achieve real-time playback for future 2160p50 content, however,
as was shown in Chapter 6. To obtain real-time playback for 2160p50 videos and
enjoy videos in even higher fidelity than today, the H.264 decoder needs to be completely parallelized, i.e., all its stages need to be parallelized.

In the previous chapters we have taken the steps understanding the application,
discovering the parallelism, mapping the parallelism onto the architecture, extracting more parallelism, and addressing the bottleneck. Now is the time to take the last
step: put together everything we have learnt in the previous chapters to implement a
highly efficient and scalable parallel H.264 decoder. For this to make sense, in this

chapter we target an application scenario and hardware platform that very likely might be available in the near future: 2160p50 video sequences and a many-core with 40 cores.

This chapter is organized as follows. In Section 7.2 we describe how the different stages of H.264 decoding can be pipelined. Between the stages buffers are needed and when buffer entries can be released is quite intricate, since the frame display order is different from the decoding order. Next, in Section 7.3, we present how the entropy decoding stage can be decoded at the frame-level. The implementation presented in this section is similar to the parallel CABAC decoder described in the previous chapter, but in this chapter we describe how it can be integrated in the over-all pipelined H.264 decoder. Thereafter, in Section 7.4 it is described how the mac-roblock (MB) reconstruction stage can be implemented in an efficient and scalable manner. The presented implementation is similar to the Ring-Line implementation presented in Section 4.4, but partially overlaps the decoding of consecutive frames in order to eliminate the ramping phases, which were a scalability limitation of the Ring-Line approach. In the article on which this chapter is based [1] the number of entropy decoding threads and the number of MB reconstruction threads had to be set statically, which is a limitation since the optimal trade-off depends on the video content as well as on the core count. In Section 7.3 we solve this problem by presenting a method to dynamically balance the number of entropy decoding and MB reconstruction threads using unified decoding threads. Experimental results are provided in Section 7.6 and conclusions are drawn in Section 7.7.

7.2 Pipelining H.264 Decoding

Besides the entropy decoding and MB reconstruction stages, there are two other stages in (our implementation of) H.264 decoding: *parse* and *output*. The *parse* stage reads the bitstream from disk and parses the high-level syntax. In particular, it detects the slice/frame markers, which is necessary to exploit frame-level par-allelism. As we did in previous chapters, we assume the worst-case one slice per frame, since it yields the highest compression ratio. The *output* stage releases buffer entries, reorders frames into display order, and displays the next frame on the screen.

Figure 7.1 depicts the high-level design of the pipelined parallel H.264 decoder. In the next section we will zoom in on the entropy decoding (labeled as ED in the figure) and the MB reconstruction (MBR) stages. The stages are connected via First-In-First-Out (FIFO) queues that contain pointers to entries in the slice syntax buffer and/or decoded picture buffer.

The *parse* stage/thread reads the input bitstream from disk and parses the high-level syntax of the bitstream. There is a single thread that performs the *parse* stage since only one thread can read from disk. In particular, the *parse* thread detects markers that signal the start of the next slice/frame, and when a new slice is detected, it allocates an entry in the slice syntax buffer (denoted by ssbe in the figure) and

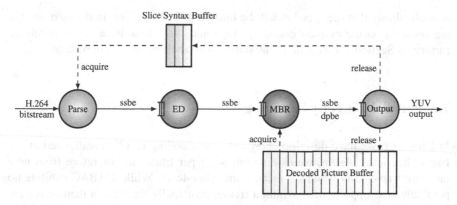

Fig. 7.1: High-level design of the pipelined parallel H.264 decoder.

then sends a pointer to this entry together with the compressed frame to the entropy decoding stage.

The *entropy decoding* stage entropy decodes using CABAC the bitstream corresponding to a frame and writes its output to a slice syntax buffer entry. As explained in the next section, there can be many entropy decoding threads processing different frames at the same time. After the output has been written to the slice syntax buffer entry, a pointer to this entry is passed on to the MB reconstruction stage.

The *MB reconstruction* stage receives pointers to slice syntax buffer entries containing, amongst others, an entropy decoded frame, and performs the decoding kernels that reconstruct the final output frame. This includes the mode prediction, intra-prediction, motion compensation, inverse transform, and deblocking filter kernels. As will be described in Section 7.4, like for the entropy decoding stage, there can be many MB reconstruction threads. Unlike entropy decoding, which is based on frame-level parallelism, the MB reconstruction stage is based on exploiting MB-level parallelism (as well as frame-level parallelism). When a MB reconstruction starts, it first allocates an entry in the decoded picture buffer. After the frame/picture have been reconstructed, pointers to the slice syntax buffer entry and the corresponding decoded picture buffer entry are sent to the output stage.

Finally (well, not finally, because the stage is repeated for the next frame), the *output* stage first marks the decoded picture buffer entries that will no longer be used as reference frames. If marked pictures are also already displayed, the corresponding decoded picture buffer entry is released so that it can be reused by another frame. (A frame can only be released if (a) it has been displayed, and (b) there are no future frames that will use it as a reference frame.) Thereafter, the corresponding slice syntax buffer entry is also released as it is no longer needed. Finally, the decoded picture buffer entry is reordered to display order and, if the next picture is available, it is written to the frame buffer so it can be displayed.

In the previous chapter we have shown that for crf22 sequences, the MB reconstruction and entropy decoding times are about the same on average. Therefore, if we only exploit pipelining, the speedup will be determined by the slowest stage and

Amdahl shows that the speedup will be limited to 2. Therefore, in the next section we zoom in on the entropy decoding stage and show how it can be parallelized further. In Section 7.4 we will do the same for the MB reconstruction stage.

7.3 Parallel Entropy Decoding

In Chapter 6 we have described how entropy decoding can be parallelized at the frame/slice-level. Again, we assume one slice per frame and therefore from now on write frame-level when we mean frame/slice-level. While CABAC exhibits no parallelization opportunities within a frame, principally the frames themselves are independent and can be entropy decoded in parallel. In order to achieve this, however, mode prediction has to be delayed until the MB reconstruction, while in most sequential implementations, it is included in the entropy decoding stage, because this results in slightly lower control overhead. Mode prediction has to be decoupled from CABAC decoding because in mode prediction there can be a dependency between a MB in a B-frame and the co-located MB in the closest reference frame. If mode prediction is moved to the MB reconstruction phase, this dependency can be covered by other dependencies between MBs and, therefore, naturally fullfilled, and it does not reduce the parallelism in the MB reconstruction stage.

Figure 7.2 shows how parallel entropy decoding can be integrated in the pipelined H.264 decoder. In this design multiple entropy decoding threads (EDTs) fetch entropy decoding tasks, in the form of a slice syntax buffer entry, from a *thread-safe* queue that is shared between the entropy decoding tasks. Thread-safe means that accesses to the queue are protected by a lock.

The number of entries in this queue as well as the slice syntax buffer has to be large enough to keep all entropy decoding threads busy. At least there should be one entry per entropy decoding task, and if the entropy decoding time varies significantly across frames, additional entries may be required.

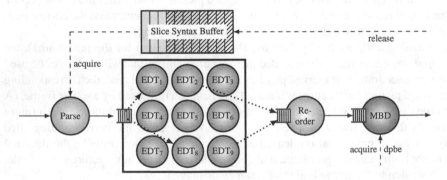

Fig. 7.2: Integrating parallel entropy decoding in the pipelined H.264 decoder.

An additional reorder thread is also necessary, because the entropy decoding threads can complete out-of-order, since the entropy decoding time varies from frame to frame. For the MB reconstruction stage, however, it is necessary to receive the frames in decoding order to avoid complications in the management of the reference frames. The reorder thread ensures that frames are sent to the MB reconstruction stage in decoding order, similar to a reorder buffer in superscalar processor organizations. The fact that there are several entropy decoding threads is completely transparent to the MB reconstruction stage.

7.4 Parallel Macroblock Reconstruction

In Chapter 4 we have presented the Ring-Line implementation of the 2D-Wave that incurs low communication and synchronization overhead, but suffers from ramping phases where the amount of parallelism is low. In Chapter 5 we have presented an implementation of the 3D-Wave that exhibits huge amounts of parallelism, but requires an efficient implementation of a subscription/kick-off mechanism which is not available on all platforms. In this section we present an implementation, called *Overlapped Ring-Line* (ORL), that combines the best of both worlds. It is similar to the Ring-Line implementation, but does not suffer from ramping phases. Viewed another way, it is also similar to the static 3D-Wave algorithm described in Section 3.3.3.2, but unlike the static 3D-Wave where the maximum motion vector length is a parameter, the ORL implementation assumes the maximum motion vector length *according to the standard*, which is 512 pixels in vertical direction and 2048 pixels in horizontal direction. Therefore, the amount of parallelism is no longer a property of the actual input video, but is guaranteed for a particular resolution.

Figure 7.3 illustrates the ORL implementation. There are several MB reconstruction threads (MRTs), and each MRT processes a whole line of MBs. Furthermore, the MB lines are distributed cyclically over the MRTs. The first MRT processes the first MB line, the second MRT the second MB line, and so on, and the nth MRT processes the nth MB line, and the $n + 1$st MB line is again processed by the first MRT, etc. Unlike the earlier Ring-Line implementation, however, the ORL implementation does not stop after processing the last MB line in a frame. It continues with the first MB line in the next frame, thereby eliminating the ramp-up/ramp-down phases. Thus at a certain time two frames are reconstructed in parallel, as illustrated in the figure. Since according to the H.264/AVC standard the maximum motion vector length in the vertical direction is 512 pixels, the dependencies between a MB and its reference areas are obeyed provided the number of MB reconstruction threads is smaller than the frame height in MB lines minus 512 (i.e., 103 for 2160p video sequences and 36 for 1080p video sequences). Actually, we also need to account for vertical interpolation, but in this description we ignore this for simplicity. (In the actual implementation we do account for vertical interpolation.)

Similar to the Ring-Line strategy, each MB reconstruction task (MRT) decodes a line of MBs. The dependencies between a MB decoded by MRT$_i$ and its upper-right

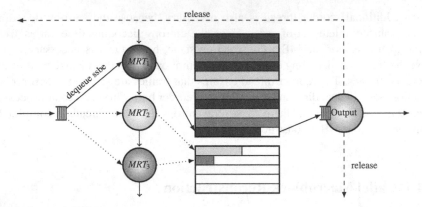

Fig. 7.3: ORL implementation of the 2D-Wave / static 3D-Wave. Ramping ineffi-ciencies are eliminated by overlapping the MB reconstruction stage of consecutive frames.

neighbor decoded by MRT_{i-1} are satisfied when MRT_i "stays behind" MRT_{i-1}. Specifically, at any time MRT_{i-1} must have processes at least two MBs more than MRT_i. The MRT that processes the last MB line in a frame informs the output thread of the completion of the frame.

Figure 7.4 illustrates how the parallel MB reconstruction stage based on the ORL strategy is integrated in the H.264 decoding pipeline. Again there is a thread-safe queue from which the pointers to slice syntax buffer entries are fetched. The MB reconstruction threads communicate and synchronize among each other in a logi-cal ring organization. In this case no reorder thread is required, because the frames are completed in-order. One additional decoded picture buffer entry is required com-pared to the Ring-Line implementation of the 2D-Wave, in order to be able to decode two frames simultaneously.

The only question remaining is: how do select the right number of entropy decod-ing and MB reconstruction threads? In the article on which this chapter is partially based [1] we used a static approach, in which the optimal trade-off is determined off-line experimentally. In the next section we improve upon this by presenting a method to dynamically balance the number of entropy decoding and MB recon-struction threads using *unified decoding threads*.

7.5 Dynamic Load Balancing using Unified Decoding Threads

The profiling results presented in Chapter 6 have shown that the execution times of the entropy decoding and MB reconstruction stages as well as their proportionality are not constant but depend strongly on the coding modes, the input video sequence, and the resolution. Therefore, it is difficult to determine a good trade-off between the

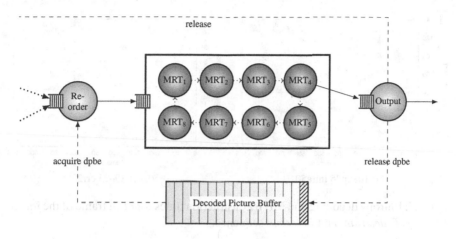

Fig. 7.4: Integrating the ORL implementation of the MB reconstruction stage in the H.264 decoding pipeline.

number of entropy decoding threads and the number of MB reconstruction threads. Perhaps reasonable trade-offs can be determined statically, off-line, by decoding many videos with different coding modes, bitrate targets, amounts of motion, etc. Such an approach would require a small database and depending on the type of video (which would have to be communicated from the encoder to the decoder), the decoder can select the number of entropy decoding and MB reconstruction threads.

Such a static approach cannot, however, resolve the load imbalance resulting from variations within a video sequence. To illustrate this problem, Figure 7.5 depicts the entropy decoding and MB reconstruction times per frame, for a high-quality video sequence encoded using intra-prediction only (Figure 7.5a) and a sequence encoded using a Constant Rate Factor of 22 (Figure 7.5b). The x-axis displays the frame number, and the y-axis the entropy decoding time and the MB reconstruction time of the frame with that number. For both encoding settings, the joined *BlueSky* (Frames 0 to 216), *Pedestrian* (Frames 217 to 591), and *Riverbed* (Frames 592 to 842) video sequence is used. Figure 7.5a shows that the entropy decoding and MB reconstruction times are fairly constant when intra-prediction only is employed, although there are some large ups and smaller downs, especially at scene changes. The CRF-22 encoding mode, on the other hand, exhibits strong fluctuations of the entropy decoding and MB reconstruction times. For such videos a static load balancing approach would be insufficient.

To solve this load balancing problem, we propose a design with *unified decoding threads* that can dynamically select an entropy decoding or MB reconstruction task to perform. Such unified threads are similar to stem cells which are able to differen-

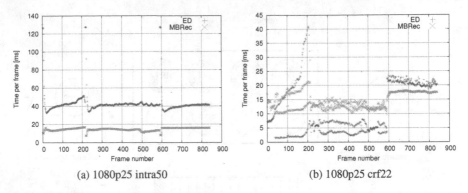

(a) 1080p25 intra50   (b) 1080p25 crf22

Fig. 7.5: Entropy decoding and MB reconstruction times for each frame of the joined *BlueSky*, *Pedestrian*, and *Riverbed* sequence.

tiate into diverse specialized cell types. Figure 7.6 illustrates the design with unified decoding threads.

In this design there are many unified decoding threads (UDTs). Each UDT can either perform an entropy decoding task or a MB reconstruction task. For example, in the figure UDT_{11} currently performs a entropy decoding task since it fetches a task from the output queue of the parse stage, while UDT_8 currently carries out a MB reconstruction task since it fetches a task from the output queue of the reorder thread and writes its results to the input queue of the output thread.

In order to synchronize the MB reconstruction tasks, a "line sync buffer" is introduced that stores some information about each UDT that executes a MB reconstruction task, such as the number of MBs it has decoded in the MB line L, and a pointer to the entry associated with the task that processes the MB line above L. To explain why this buffer is needed, please refer back to Figure 4.6 on Page 44 that sketches the Ring-Line implementation on the Cell processor. Due to the vertical MB dependencies, cores with consecutive IDs need to communicate and synchronize with each other. There it was relatively simple, because only the MB reconstruction stage was considered. When there are unified decoding threads, however, and a core can switch between an entropy decoding and a MB reconstruction task, there is no direct way of knowing which core processes which MB line. To solve this problem, the line sync buffer is used, which is a data structure stored in shared memory, and each UDT can find in this buffer which UDT processes the previous MB line and which UDT processes the next MB line, as well as some other information. In other words, the UDTs performing MB reconstruction tasks are linked together using this data structure.

With two task queues (the output queues of the parse stage and the reorder thread), a new challenge is introduced: from which queue should a UDT fetch a new task? The task selection policy should incur low overhead, achieve good load balancing, and be scalable. These requirements can be achieved by using a sim-

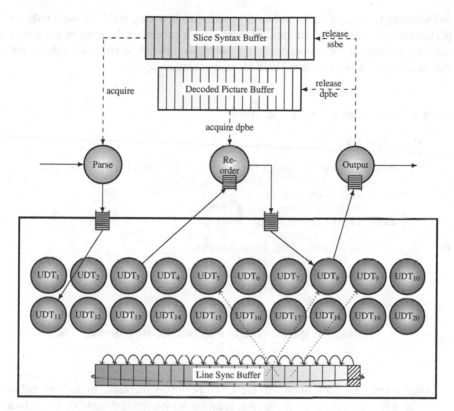

Fig. 7.6: Load balance regardless of the input sequence can be achieved using unified decoding threads.

ple *"entropy-first"* policy. As the name indicates, each unified decoding thread first checks if an entropy decoding task is available by inspecting the output queue of the parse stage. If so, it executes this task. If not, it fetches a MB reconstruction task from the output queue of the reorder thread. This makes sense because there is no work to do in the MB reconstruction stage if frames have not been entropy decoded first. The frame latency is kept under control by having several but not too many slice syntax buffer entries. This ensures that when several frames have been entropy decoded, they need to be reconstructed and displayed first to make room for additional frames.

To explain the entropy-decoding first policy and the concept of unified decoding threads in somewhat more detail, some high-level pseudo-code is presented in a top-down fashion. Listing 7.1 depicts pseudo-code for a unified decoding thread. It executes a loop until a shared-memory variable `finished` is set that signals that the entire video sequence has been decoded. It each loop iteration the UDT first gets a new task to perform. It does so by calling `get_new_task` until `get_new_task` returns a task, a process known as *spinning*. In the actual implementation we do not use spinning because it consumes more power than blocking and because it reduces performance in case there are multiple threads per processor. Instead, in the actual implementation we use a sleep/wake-up mechanism based on so-called *tokens*, but this mechanism is not presented here for understandability.

```
void unified_decoding_thread()
{
    while(!finished)
    {
        while ((task = get_next_task()) == NULL) /* spin */ ;

        if (task->type == ED)
            decode_entropy_frame(task);
        else if (task->type == MBR)
        {
            reconstruct_mb_line(task);
            if (task->last_line)
                enqueue(output_q, task);
        }
    }
}
```

Listing 7.1: Pseudo-code of a unified decoding thread.

After a task has been obtained, the UDT performs the task, by calling either decode_entropy_frame if the task is an entropy decoding task, or by calling reconstruct_mb_line if it is a MB reconstruction task. Furthermore, if the MB reconstruction task is the task that processes the last MB line in a frame, the task is enqueued to the input queue of the output thread.

Listing 7.2 presents pseudo-code for the function get_next_task that implements the entropy-decoding first policy. It first checks the queue containing entropy decoding tasks (the output queue of the parse thread). If it is not empty, it dequeues the first task. The queues are shared data structures that can be read and written by several threads simultaneously and therefore need to be protected by locks.

If a task has been obtained from the queue of entropy decoding tasks, it is immediately returned. Otherwise the queue of MB reconstruction tasks is checked, and if it is not empty, the first task in this queue is dequeued. Again accesses to this shared queue need to be protected by locks. Finally, get_next_task returns the task pointer, which either points to a task structure if a task has been obtained from the queue of MB reconstruction tasks, or is NULL if no task has been obtained from either task queue. We note again that the actual implementation is more intricate, since it does not use spinning and since we have not precisely described how entries in the line sync buffer are allocated and linked together.

Finally, Listing 7.3 shows pseudo-code for the reconstruct_mb_line function that implements a MB reconstruction task. It first gets the line sync buffer entry (lsbe) associated with the task which is kept in the structure that represents the

```
TASK get_next_task()
{
    task = NULL;

    lock(entropy_q);
    if (!is_empty(entropy_q))
        task = dequeue(entropy_q);
    unlock(entropy_q);

    if (task != NULL)
        return task;

    lock(mbr_q);
    if (!is_empty(mbr_q))
        task = dequeue(mbr_q);
    unlock(mbr_q);

    return(task);
}
```

Listing 7.2: The get_next_task function implementing the entropy-first job selection policy.

task. From this buffer entry it gets the line sync buffer entry associated with the task that reconstructs the previous MB line (lsbe_prev). Then for each MB in the MB line that it reconstructs, it first waits to ensure that the task that reconstructs the previous MB line has processed at least 2 MBs more than itself. Here we see a glimpse of the convenience of shared-memory programming, since this waiting can be relatively easily implemented by spinning on two shared variables that are obtained by dereferencing pointers. Thereafter, the current MB is decoded and the counter field of the lsbe structure representing the number of MBs that have been processed is incremented to signal to the MB reconstruction task that decodes the next line that it can continue. A small technical detail is that when all MBs in the MB line have been processed, the counter has to be incremented once more to prevent deadlock.

7.6 Experimental Evaluation

To evaluate the parallel H.264 decoder presented in this chapter, we have implemented it using the sequential H.264 decoder of the FFmpeg project [2] as the starting point. The FFmpeg H.264 decoder is heavily optimized for various processor architectures and is one of the fastest *sequential* H.264 decoders available. Among others, key kernels of the H.264 decoder have been implemented using the MMX/SSE SIMD extensions of the x86 architecture. As input video sequences we used the sequences described in Section 6.2 on Page 69. As described in detail there, these sequences have been obtained by joining two or three videos to emulate real

```
void reconstruct_mb_line( &task )
{
    lsbe = task->lsbe;
    lsbe_prev = lsbe->prev;

    for (each mb in task->line)
    {
        while (lsbe->mb_processed < lsbe_prev->mb_processed-1) /*
            spin */ ;
        decode_mb();
        lsbe->mb_processed++;
    }
    lsbe->mb_processed++;    // additional incrementation to
        finish line
}
```

Listing 7.3: Decode macroblock line with wavefont synchronization.

videos with different scenes. The sequences have been encoded using the constant quality encoding mode with a constant rate factor or intra-only mode. CRF-22 is representative of high quality, high definition applications, and the other extreme, CRF-37, is typical for medium- to low-quality Internet video streaming applications (e.g., youtube). Intra-only encoding is representative of professional video systems such as video editing and high quality camera equipment. This mode uses a constant bitrate with 50Mbps for 1080p25 and 400Mbps for 2160p50.

The evaluation platform is a Westmere-EX cc-NUMA system consisting of four sockets, each containing a 10-core processor. Each core supports simultaneous multithreading (SMT) with 2 threads per core. The SIMD unit is shared between the 2 threads. While such high-performance server systems are presently not used to decode videos, it is likely that in the not so distant future desktop systems will provide similar performance levels and core counts. More details about the evaluation platform can be found in Table 6.2 in Section 6.4 on Page 76.

Results are presented for 1 to 40 cores in steps of 5 cores. Experiments have been conducted with one thread per core (indicated by *smt off* in the figures) as well as two threads per core (*smt on*). For the *smt off* case one unified decoding thread is spawned on each core, while for *smt on* two threads are launched per core. To reduce the influence of the OS on the timing results, each thread is *pinned* to a core. Pinning means that a thread is binded to a specific core, which is achieved by using a Linux system call `sched_setaffinity` to change the cpu mask of the particular thread.

Figure 7.7 depicts the speedup as well as the throughput in frames per second (fps) as a function of the number of cores. Many conclusions can be drawn from these results. First and foremost, the presented parallel H.264 decoder achieves very high performance, ranging from 410 fps to 861 fps for the 1080p25 sequences and from 99 fps to 267 fps for the 2160p50 sequences. These throughputs exceed the performance required for real-time playback (25 fps for 1080p25 and 50 fps for

2160p50) by far. Obviously, the higher the resolution, the lower the throughput in fps. For 1080p25 real-time performance is achieved on a single core, but for 2160p50 about 10 cores are required to achieve real-time performance for each encoding mode. The particular speedup depends on the encoding mode used to encode the video sequences.

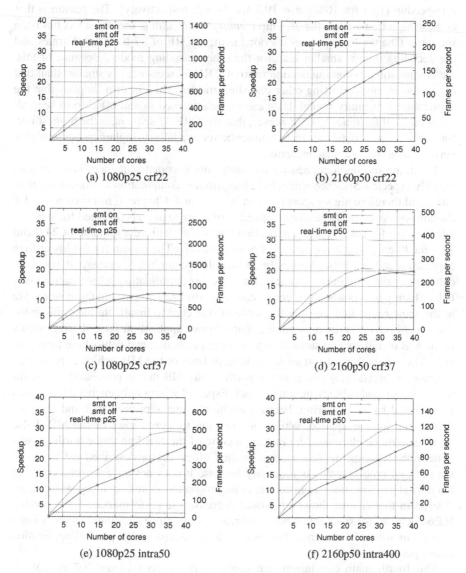

(a) 1080p25 crf22

(b) 2160p50 crf22

(c) 1080p25 crf37

(d) 2160p50 crf37

(e) 1080p25 intra50

(f) 2160p50 intra400

Fig. 7.7: Scalability (speedup) of the presented parallel H.264 decoder on a 40-core cc-NUMA architecture.

Second, although the performance in fps is very high, the speedup in most cases is less than linear, especially for the bitstreams encoded with a CRF of 37. For example, with 1 thread per core (smt off), the maximum speedup for the 1080p sequence is 24.0 and for the 2160p sequence it is 27.9. In general, and not surprisingly, the speedup for the 2160p sequences is higher than for the 1080p sequences, since they exhibit more parallelism. Sequences encoded with a CRF of 37 achieve the highest throughput (861 fps (smt off) for 1080p and 254 fps for 2160p) but also the lowest speedup (12.4 for 1080p and 19.8 for 2160p, respectively). The reason is that at these high frame rates, the off-chip *memory bandwidth is saturated*. As has been shown in Chapter 6, sequences encoded using a CRF of 37 have low bitrates and the decoding time is smaller than for the other encoding modes. Because of this, the computation-to-communication ratio of CRF-37 sequences is smaller than that of sequences encoded using other encoding modes. This, in turn, implies that the application is memory-bound. Data cannot be streamed in and off chip to keep all cores busy all the time. Note, however, that for CRF-37 sequences the highest performance in fps is obtained. Just as more=better is a fallacy, higher speedup=higher performance is also a misconception.

Third, the speedup obtained for the intra-only encoded sequences is lower than initially expected. Such sequences have a significant computation-to-communication ratio, and therefore higher speedups can be expected. Chapter 6 has shown that for the entropy decoding stage alone a speedup of more than 36 is obtained for such sequences, but for the complete application we "only" achieve a speedup of 24.0 and 25.3 on 40 cores for 1080p and 2160p, respectively. The reason for the lower than expected speedup is *false sharing*. In cache-coherent multicore systems, data is kept coherent at the granularity of cache lines, i.e., by invalidating whole cache lines. False sharing occurs when two cores read/write different words, but these words are in the same cache line, causing the write of one core to invalidate the (still up-to-date) word read or written by another core. In our application false sharing occurs in the MB reconstruction stage during the intra prediction and deblocking filter kernels. These kernels read and write the bottom lines of the MB (which is processed by another thread/core) that is to the north of the MB that is processed by the intra prediction and deblocking filter kernel. Especially intra only encoded sequences are affected by false sharing, because all blocks are intra predicted and a strong deblocking filter is applied, which modifies the three top rows of the MB. False sharing could be eliminated by making a copy of the MB line to the north, as was done in the Ringline implementation on the Cell processor described in Chapter 4. On the Cell processor this is necessary because it does not have a cache-coherent but a local store memory architecture. Doing so on the shared-memory system considered in this chapter, however, would decrease the performance of the parallel H.264 decoder for smaller numbers of threads/cores, since additional memory copy operations would be required. This is a trade-off decision that parallel application developers have to take.

The fourth main conclusion that can be drawn from Figure 7.7 is that although turning simultaneous multithreading (SMT) on improves performance in most cases, the improvements are much smaller than might be expected. In the *ideal* world, since in this case there are 2 threads per core, the performance improves by a factor of 2. The graphs show that the obtained improvements are much smaller. Up to 25 cores, SMT on always improves performance (by at most 1.47). For more cores, the performance of SMT on is sometimes even lower than when SMT is turned off. The reason for this behavior is that, although each core can execute two threads simultaneously, these threads need to share most processor resources such as Arithmetic Logic Units (ALUs), floating-point units, and load/store queues. The intra-only sequences in general profit more from SMT than the CRF sequences, because intra-only sequences spend relatively more time on MB reconstruction than on entropy decoding. Furthermore, since the MB reconstruction kernels heavily employ SIMD instructions while CABAC decoding does not, intra-only sequences suffer more from the fact that there is only one SIMD unit per core. Did we already identify the step *"understand the architecture"* that developers of multi-core software also have to take? If not, it is because we would need a bigger book.

We could continue discussing and analyzing the results, but it would distract from the main message. After taking a number of design steps, we have developed a highly efficient and scalable parallel H.264 decoder. The fact that it does not scale perfectly on the particular evaluation platform used in this chapter is *not* due to a scalability limitation of the implementation, but due to a scalability bottleneck of the platform that we recently purchased.

7.7 Conclusions

In this chapter we have taken the last step in our informal parallel-application-design-process: put it all together. This step might also be called "re-iterate", "do it all over again", or "do it right this time". We have presented a high-performance, fully parallel H.264 decoder that achieves faster than real-time performance for high-resolution 2160p50 video sequences. One novel aspect of this implementation that has not been published before is that it employs unified decoding threads which, like stem cells, can either perform an entropy decoding task or a MB reconstruction task. This improves performance if the time taken by the entropy decoding and MB reconstruction tasks varies significantly, as is the case for H.264/AVC video sequences.

Two main conclusions can be drawn from this chapter. The first is: *it's the memory bandwidth bottleneck!* Although the obtained performance (fps) is very high and more than sufficient, the speedup is less than linear because the data cannot be streamed onto and off chip to keep all cores busy all of the time. This is a major concern for parallel application developers as well as computer architects. Our second main conclusion is that parallel application developers also need to *understand the architecture* in order to be able to develop highly-efficient parallel applications. In particular, it has been shown that architecture features such as false sharing and functional unit sharing have to be taken into account when developing high-performance parallel software.

In the next chapter we conclude this book by briefly summarizing its contents, drawing our main conclusions, and by having a brief look at what the future might have to offer.

References

1. Chi, C.C., Juurlink, B.: A QHD-capable Parallel H.264 Decoder. In: Proceedings of the International Conference on Supercomputing, pp. 317–326 (2011)
2. FFmpeg. http://ffmpeg.org

Chapter 8
Conclusions

We have come to the end of this book. We hope you enjoyed reading it as much as we liked writing it (well, there were times ..., we hope that you did not have similar times). In this final chapter we briefly summarize each chapter, draw our final conclusions, and have a brief look at the parallelization opportunities in the upcoming video standard HEVC.

The first step that any programmer who wants or needs to parallelize an application has to perform is to understand the application. There might be applications where 99.9% of the execution time is spent in a small kernel consisting of at most a few dozen lines of code, allowing the developer to focus on this kernel, Unfortunately, this is certainly not the case for H.264 decoding. Therefore, in Chapter 2 we have presented a brief overview of the H.264 standard. The goal of this chapter was not to present all the details of the H.264 standard, but to provide sufficient information to understand the remaining chapters.

The second step that needs to be followed is to discover or expose the parallelism in the application. In Chapter 3 we therefore discussed various ways to parallelize H.264 decoding. The two most promising approaches, both based on exploiting parallelism between different macroblocks (MBs), were evaluated in more detail using high-level simulation, assuming that the time to decode a MB is constant and that there is no communication and synchronization overhead. These analytical results show that the first approach, referred to as the 2D-Wave, should be able to scale to about 20-25 cores for Full High Definition (1080p) video sequences. Furthermore, the parallelism scales with the resolution, so the scalability of the 2D-Wave increases with the resolution. The second approach, referred to as the 3D-Wave, achieves much higher scalability. We found that using this approach, up to about 9000 MBs could be processed in parallel, which is of course a huge number. We concluded that there is sufficient parallelism in H.264 decoding to make it scale to a many-core architecture containing 100 cores and beyond.

Discovering the parallelism, however, is not enough. In the third step the parallel programmer also has to find efficient ways to exploit the parallelism. In other words, he or she has to map the parallelism onto the system. In Chapter 4 we presented two different ways of implementing the 2D-Wave algorithm on the Local Store architecture of the Cell processor. The first implementation, referred to as the Task Pool (TP), uses a master core (the PPE) to keep track of the dependencies between MBs and maintains a pool of MB decoding tasks that are ready to be executed. Slave cores (the SPEs) continuously request a task from the master, execute the task, and then inform the master when they are done and request a new task. This implementation has the advantage that it can achieve perfect load balancing, because all the work that can be performed is kept in a centralized task pool. In the second implementation, called the Ring-Line (RL) implementation, each SPE processes whole lines of MBs. This has the disadvantage that it may cause load imbalance, but has several other advantages. First, cores do not have to synchronize via the central master (which may become a bottleneck) but can synchronize point-to-point. Second, because it is known which core will process which MB, intermediate results do not have to be communicated via main memory but can be communicated on chip. Third, for the same reason, data can be prefetched and communication can be overlapped with computation, allowing the DMA latency to be hidden. Indeed, the experimental results showed that while the TP implementation achieves better scalability in theory, in practice the RL implementation attains much higher performance. This shows that for some applications, issues such as load balancing and communication/synchronization cost need to be balanced against each other. In our experience in most cases what works well on one multicore-architecture (balance the load, reduce communication and synchronization overhead, hide memory latency, etc.), also works well on other architectures. In highly dynamic and irregular applications such as H.264 decoding, however, this is not the case and a trade-off needs to be made.

The implementations presented in Chapter 4 achieve (faster than) real-time performance for 1080p25 applications, but do not scale to a many-core system containing 100 cores or more. Therefore, if an application requires higher performance than is provided by the initial implementation, the next step application developers have to take is to extract more parallelism. This step was described in Chapter 5, where we presented a highly-scalable implementation of the dynamic 3D-Wave algorithm, which, as previous analytical results had shown, is able to scale to a huge number of cores. In this chapter we focused on cache-coherent shared-memory systems, and therefore we also described a shared-memory implementation of the 2D-Wave. This implementation is based on a task pool, as the TP implementation presented in Chapter 4, but unlike the implementation in Chapter 4, there is no central master core that manages the task pool. Instead, the task pool is a shared data structure in shared memory, and accesses to this data structure must be protected by atomic operations. Furthermore, in order to implement the 3D-Wave, we needed to design and implement an intricate subscription/kick-off mechanism in which MB decoding tasks can subscribe themselves to the reference MBs on which they depend. In essence, the developed mechanism is similar to the suspend/resume mechanism as provided by the conventional synchronization primitive semaphore, but the functionality required by the dynamic 3D-Wave algorithm is not identical to the functionality provided by semaphores. In our opinion this shows that sometimes developers of highly-efficient, massively parallel applications need to design their own, application specific, synchronization primitives (or future parallel programming models should provide richer sets of synchronization primitives). Experimental results obtained using a detailed simulator of a shared-memory many-core architecture show that the 3D-Wave implementation is able to scale to at least 64 cores (more cores could sadly not be simulated) and that the parallelism is not nearly exhausted at 64 cores. The 2D-Wave implementation, on the other hand, stops scaling after 16 cores.

In Chapter 4 and Chapter 5 we cheated a little. If results look too good to be true, they usually are. While the MB reconstruction phase, on which we focused in these chapters, is certainly the most time-consuming phase of H.264 decoding, there is another phase that takes a significant amount of time: the entropy decoding phase. Cheating is not the right word, since we wrote in the small print that we assumed that the entropy decoding phase would be performed by a hardware accelerator, so in court we would be safe. In Chapter 6 we analyzed the validity of this assumption, and showed that while entropy decoding can be accelerated by a hardware unit, the speedup would not be sufficient to defeat Amdahl's law. Therefore, one also has to exploit parallelism in the entropy decoding phase, and we showed that this can be done by exploiting pipelining parallelism to entropy decode different frames in parallel. We call this step "addressing the bottleneck", but state that addressing the bottleneck is a step that needs to be taken at every step of the design process, not only after parallelizing the most time-consuming phase.

The final step is to take everything we learned in the process to develop a highly-scalable, highly-efficient parallel implementation of H.264 decoding that squeezes out every cycle of performance from the application. We call this step "re-iterate", "do it all over again", or "put it all together". As usual in life, once you have done something, you know how it should have been done. In industry this step may be unaffordable, but we could exploit our so-called "academic freedom" to do it right this time (if you, the reader, work in industry, please believe us that there is much less academic freedom than it may appear from the outside). In Chapter 7 we, therefore, presented a parallel implementation of H.264 decoding that includes all phases but still scales to a large number of cores. The implementation exploits pipelining as well as data-level parallelism and its MB reconstruction phase is based on the "static" 3D-Wave parallelization strategy that assumes the worst-case motion vector length (512 pixels in level 4.0 and beyond of the H.264 standard). Due to the fact that the vertical motion vector is at most 512 pixels (for 1080p resolution), the MB reconstruction phase of two consecutive frames can be partially overlapped, which gets rid of the "ramping" effects of the 2D-Wave. The thus obtained parallelism is more than sufficient for real-time decoding and very fast forwarding and, furthermore, scales with the resolution. The experimental results obtained on a 40-core cc-NUMA architecture show that the presented implementation achieves very high performance, from 99 frames per second (fps) to 267 fps for high-resolution 2160p50 sequences. For real-time playback, "only" 50 fps is required.

We hope we have managed to convince you that, currently, scalable parallel programming is challenging and an art. However, because it is challenging, it is also fun. We strongly believe, however, that this situation cannot remain (the fun part yes, the complexity part not). The junior members of the author team claim that parallel programming is not so hard, but the senior members conjecture that this is because they have educated them well. As this book has shown, parallel programmers not only have to discover and express the parallelism, they also have to take care of load balancing, reducing communication and synchronization cost, hiding memory latency, scheduling, protecting accesses to shared data structures (atomic operations), data locality (tail submit), implementing new synchronization primitives if the supported ones do not provide the required functionality, struggle with dependencies that are only there due to legacy issues, unify different types of tasks and then deal with the synchronization problem it introduces, to name just a few issues. Again we would like to emphasize that it is not out intention to scare away people from parallel programming. On the contrary, by presenting a parallel-application-design-process we hope to interest more people in the art of parallel programming.

If multi-/many-core programming is to become mainstream, however, or at least is to play a much larger role than today, many of these programming issues have to be hidden as much as possible from the programmer. This might be done by, for example, providing higher-level parallel programming models, better tool support, and, if possible, support in the multi-/many-core architecture to relieve the parallel programming burden. Some of these issues are addressed in the European project ENCORE (Enabling Technologies for a Programmable Many-Core) [2], in which some of the authors are involved. In the ENCORE project, a novel parallel programming model is being developed and evaluated in which the programmer can express parallelism by annotating pieces of code (typically functions) as tasks as well as the inputs and outputs of those tasks. Based on the inputs and outputs of each task, among other things, the run-time system can determine the dependencies between tasks, schedule ready tasks onto available cores, and optimize data locality. The parallel programmer, therefore, does not have to worry about about low level issues such as synchronization and optimizing data locality, but can focus on discovering the parallelism instead. In the ENCORE project, we have developed an implementation of H.264 decoding using the ENCORE programming model. Unfortunately, the number of pages allowed in this book was too small to include a description of the ENCORE programming model and to present the results of this implementation. Determining and keeping track of the dependencies between tasks, however, can take a lot of time which can limit the scalability of the ENCORE approach. Therefore, another goal of the ENCORE project is to develop hardware support for this task, in line with the step "addressing the bottleneck".

Parallelism in the H.264 standard was an after-thought. The main goals of the H.264 standard were high compression and high visual quality. Real-time execution was expected assuming that single-core performance would continue to increase as before and that hardware solutions could be developed. Treating parallelism as an after-thought has to change in the future, since single-core performance has stopped increasing and developing hardware is time-consuming and hence costly. Interestingly, the upcoming video coding standard HEVC (High Efficiency Video Codec), which is currently being developed, includes proposals specifically aimed at parallelism. Some of these proposals are named tiles, entropy slices, and Wavefront

Parallel Processing. The general idea behind all of them is to allow the creation of picture partitions that can be processed in parallel without incurring high coding losses. Picture partitions solve a limitation of the implementation presented in Chapter 7, namely the frame buffer that is needed between the entropy decoding and macroblock reconstruction stages. This buffer increases the memory footprint of the application and decreases data locality, which has a negative impact on performance and power. With the new picture partition approaches in HEVC, it is possible to also decrease frame latency, which is an important issue for video conferencing applications.

We have already begun to evaluate these proposals as well as to propose modifications to these proposals to make them more efficient, but again, this book is too short to include these results. The interested reader is referred to our publications [1]. To be continued ...

References

1. Alvarez-Mesa, M., Chi, C.C., Juurlink, B., George, V., Schierl, T.: Parallel Video Decoding in the Emerging HEVC Standard. In: Proceedings of the 37th International Conference on Acoustics, Speech, and Signal Processing (ICASSP) 2012 (2012)
2. Encore Project: ENabling technologies for a future many-CORE. URL http://www.encore-project.eu/